Kevin
13 Stowe Rd.
PORTSMOUTH.
ENGLAND.

KUMAUN
THE LAND AND PEOPLE

KUMAUN
THE LAND AND PEOPLE

S. S. NEGI
Deputy Conservator of Forests

INDUS PUBLISHING COMPANY
NEW DELHI

First published in 1993
by Indus Publishing Company
FS-5, Tagore Garden, New Delhi 110027

ISBN 81-85182-89-2

Published by M. L. Gidwani, Indus Publishing Company
FS-5, Tagore Garden, New Delhi 110027, and printed at
Efficient Offset Printers, Shahzada Bagh, New Delhi 110035

Preface

Kumaun region of the Himalaya forms a part of the State of Uttar Pradesh. The borders of Kumaun touch those of Tibet, Nepal, Garhwal and the plains of western Uttar Pradesh. The entire tract is mountainous with low Siwalik hills in the south and the gigantic main Himalayan mountain wall in the north.

This beautiful hilly region is made up of administrative districts of Nainital, Almora and Pithoragarh. The mountains are clothed with dense forests ranging from sal forests in the south to alpine meadows just below the snowline in the north. Kumaun is a land of snow-clad mountain peaks, fast-flowing waters, placid lakes, terraced fields clinging to steep mountain slopes and small villages.

This book attempts to present in a concise form the land and people of Kumaun. It deals with regional geography, climate, soils, rivers, lakes and glaciers, geology, natural vegetation and forests, wildlife and sanctuaries, history, people and culture, economy, environmental degradation, and places of interest in Kumaun.

I am indebted to Dr. D. N. Tewari, Director General, ICFRE and Shri N. K. Joshi, Director, Forest Research Institute, Dehradun for constant guidance and encouragement.

Thanks are also due to my wife, Manju, for her self-denial and to the publishers for bringing out this volume in a short time.

SHARAD SINGH NEGI

Contents

1
Regional Geography

Kumaun occupies the easternmost part of this mountainous region, lying at the trijunction of India, Nepal and Tibet. To the north of Kumaun lies Tibet which is separated by a natural water divide of the main Himalayan mountain wall. In the east the Kali Ganga separates Kumaun from Nepal, while distinct ridges separate Kumaun from Pauri and Chamoli districts of Garhwal. In the south the terai and bhabar tract of Nainital district touches the boundaries of Bijnore, Moradabad, Rampur, Bareilly and Pilibhit districts.

Three districts constitute Kumaun viz. Almora, Nainital and Pithoragarh. They are divided into 14 tehsils and 41 development blocks. Kumaun has a total geographical area of 21,035 sq. kms which accounts for about 14% of the total area of U.P. and 41% of that of the U.P. Himalaya (known as Uttaranchal or Uttarakhand). This region exhibits a socio-cultural unity amidst a physical diversity.

The altitudinal range of Kumaun varies from 200 mts in the terai and bhabar tract to over 7500 mts in the main Himalaya.

PHYSIOGRAPHIC DIVISIONS

The Kumaun Himalaya can aptly be divided into the following physiographic divisions:

1. Sub-Montane Zone

This is a gently sloping southernmost fringe or tract of Kumaun that lies between the Siwalik hills in the north and the Ganga plains of U. P. in the south. The sub-montane zone may in turn be divided into the following:

1. *Bhabar*

This is a gently sloping tract to the south of the Siwalik hills. It consists of huge boulders that have been brought down by the torrents emanating from the southern slopes of the Siwaliks. Rivers and streams may flow below the surface in the bhabar tract. The presence of boulders of varying sizes in the soil have rendered the land unfit for cultivation.

2. *Terai*

South of the bhabar tract lies the terai which is more or less flat. It forms the transition zone between the Ganga plain in the south and the Siwalik hills in the north. The water-table is very high in the terai tract. The entire area was once a marshy land which has now been reclaimed for agriculture.

A number of rivers, streams and seasonal torrents flow in from the north and drain the bhabar and terai tracts. These include the Kosi, Phika, Dhela, Baur, Dabka, Bhakhra, Gaula, Sukhi and Kali rivers. They are in spate during the rainy season. The rain water converts small depressions in the terai into swamps and marshes.

2. Siwalik Hills or Outer Himalayan Zone

The Siwalik hills or outer Himalayan zone lies to the north of the terai and bhabar tract. This may further be sub-divided into the following:

1. *Siwalik Hills*

The low rolling Siwalik hills extend along a NW-SE direction in southern Kumaun. They rise above the terai and bhabar tract. The average elevation varies from 700 to 900 mts, while the tops of the ridges may be upto 1300 mts in elevation. These hills are made up of sediments that were probably laid down during the late Tertiary or Quaternary ages.

The southern slopes of the Siwalik hills are steeper than their northern counterparts. Thick forests occur on the latter slopes while the former bear open forests and are broken by steep scarps. Numerous rivers, streams and seasonal streams drain the Siwalik hills. Sots are seasonal streams which remain dry for about nine months in a year but are in spate during the monsoon season.

2. *Dun Valleys*

There occur longitudinal valleys between the Siwalik hills in the south and the lower Himalaya in the north. They are known as dun valleys, e.g. Patli dun and Kairda dun valleys. The dun valleys are filled with recent to sub-recent sediments in the form of coarse and fine gravels that have been brought down by rivers and streams both from the north and from the south.

The dun valleys have a length varying from 20 to 40 kms and width from 5 to 15 kms. Thick forests cover these valleys. Many areas have been cleared for agriculture as the alluvial soil is fertile.

A number of rivers and streams drain the outer Himalayan zone. These include the Ramganga, Kosi, Baur, Dabka, Nihal, Nandhaur and Kali.

3. Lower Himalayan Zone

The lower Himalayan zone lies to the north of the dun valleys or Siwalik hills. It consists of the foothills of the main Himalayan range with average elevation between 1000 to 2500 mts. This zone has a width of about 60 to 80 kms.

The lower Himalayan zone may be sub-divided into two:

1. *Lower Himalaya*

The lower Himalaya are the southernmost unit of this zone. They are made up of a series of hills whose tops usually do not rise to an elevation of more than 2300 mts. The south facing slopes are steeper than their northern counterparts. Dense forests cover the north facing slopes.

2. *Middle Himalaya*

The middle Himalaya lie just to the south of the main Himalayan mountain wall. They consist of a series of hills whose tops may rise to an elevation of about 3000 mts. The southern slopes are steeper than the north facing ones. However, sparse vegetation covers the former slopes.

Many rivers and streams have carved valleys in the lower Himalayan zone. These rivers have cut across both the lower and middle Himalaya. Amongst these water channels are the Ramganga, Someshwar, Kosi, Katyur, Chhana, Saryu, Salam, Panar, Thal, Sor, Sira, Kalikumaun and Kali rivers.

4. Main Himalayan Zone

The main Himalayan zone occupies the northern tract of Kumaun. It is in the form of a mountain wall whose tops are under a perpetual cover of snow. The highest peaks soar to an elevation of over 7000 mts.

The salient characteristics of this zone are outlined below:

1. A number of high peaks are a part of the main Himalaya, e.g. Panchchuli (6994 mts), Nandadevi (7816 mts), Nandakot (6661 mts) and Trishul (7120 mts).

2. There occur glaciers in the upper valleys of the main Himalayan mountain wall. Amongst these are the Pindari, Milam and Sunderdhunga.

3. Many rivers draining Garhwal and Kumaun rise from this zone, e.g. Pindar, Gauri, Dhauli and Kali. These rivers have covered deep V-shaped valleys.

4. Forests varying from temperate to alpine cover the lower and middle slopes of the main Himalaya. The upper slopes are devoid of a vegetative cover as they are under a permanent cover of snow.

5. There occur dry valleys within this zone. These valleys receive very little rainfall as they lie in the rain shadow of the towering peaks of the main Himalayan mountain wall. Such valleys are termed as inner dry valleys.

5. Trans-Himalayan Zone

In the extreme northern tract of Kumaun is the trans-Himalaya made up of valleys lying across the main Himalayan range. This is largely a cold and dry tract as rainfall is very low. The moisture bearing SW monsoons are unable to cross the towering main Himalaya and thus this tract receives very less rainfall.

The average elevation of this tract is about 3000 mts. Vegetative cover is extremely sparse. The valleys are U-shaped and have been covered by the action of glaciers. The trans-Himalayan valleys of Kumaun are— (a) the upper valley of the Gori river (Malla Johar), (b) the upper valley of the Dhauli river (Darma and Chaudan), (c) the upper Kuti (Byans) valley.

The Unta Dhura ridge divides the catchments of the Gori or Gauri river in the south-west and the Girthi in the north. The north facing slopes of Unta Dhura ridge lie in the trans-Himalayan range.

CLIMATE

Kumaun enjoys a varied climate ranging from hot and moist tropical in the terai and bhabar tract to arctic or polar in the upper reaches of the main Himalayan range. In winter the mean temperature in the terai and bhabar is 21°C; about 8°C in the middle Himalaya, 6°C in the valleys of the higher Himalaya and below the freezing point above an elevation of about 5000 mts.

The hottest areas are in the southern hills which lie adjacent to the plains of western Uttar Pradesh. The average temperature decreases towards north though the valleys experience sub-tropical climate. The SW monsoons account for bulk of the total annual precipitation. Winter rains too cause widespread precipitation in most parts of Kumaun. Snowfall is received in the high hills.

The climate of Kumaun has been discussed in detail in a separate chapter.

NATURAL VEGETATION

The Kumaun hills support a diverse natural vegetation. These include the following forests:

1. Moist, dry, Siwalik, dun and terai-bharbar sal forests,
2. Riverine or khair and sisham forest,
3. Lower and upper mixed scrub,
4. Chir pine forest (Siwalik and lower Himalaya),
5. Temperate coniferous forests (deodar, blue pine, fir and spruce),
6. Oak forests (ban, kharsu and moru),
7. Temperate and sub-alpine deciduous forests (Rhododendron, Acer, Mapple etc.),
8. Sub-alpine forests,
9. Moist and dry alpine scrub.

The nature and composition of the natural vegetation varies with aspect, slope, altitude and climate. The natural vegetation and forests of Kumaun have been discussed in a separate chapter.

HYDROGRAPHY

1. *Permanent Snow and Glaciers*

The upper and middle slopes of the main Himalayan mountain wall is under a permanent cover of snow. The level at which the snowline is found in Kumaun is given below:

Situation/aspect	*Snowline (in mts)*
Southern and external ranges	4800
Indo-Tibetan water divide	5500
Trans-Himalayan slopes	4700

This permanent snow cap is the source of countless channels of water which combine to form the main river systems draining Kumaun. There also occur several glaciers on the slopes of the main Himalayan mountain wall. Amongst these the Pindari glacier is the largest. Other include the Milam, Talkot, Kalaba, Bamlas, Poting and Balatai glaciers.

2. *Rivers and Streams*

Kumaun is drained by many large and small rivers. These are made up of three drainage systems, viz.

a) the Alakanada river system,
b) the Ramganga river system,
c) the Kali or Kaliganga river system.

3. *Lakes*

Kumaun is also known as the land of lakes. There are many lakes in Nainital area, the largest being the Nainital lake which has a surface area of about 4.5 sq. kms. The Bhimtal lake has an area of about 3.6 sq. kms while the surface area of the Naukuchiatal lake is about 2.6 sq. kms.

There are many other smaller lakes in Kumaun. Many of these are of tectonic or glacial origin.

4. *Ground Water Resources*

Geological, climatic and topographic factors have a bearing on the ground water resources of Kumaun. A number of rivers and streams are fed by springs and seepages emanating from the ground water resources, e.g. Ramganga, Kosi, Charma, Panar, Ladhiya and

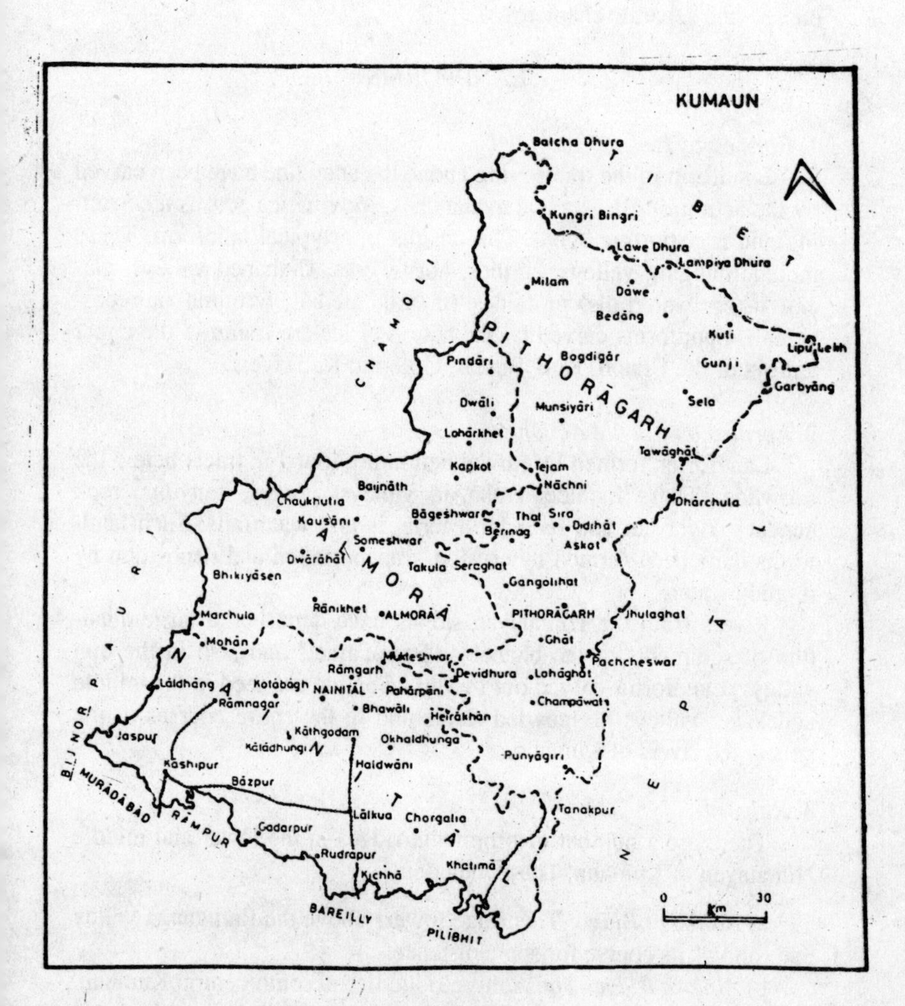

Map of Kumaun.

Gaula. Smaller tributaries of snowfed rivers are also charged by the reservoir of underground water.

Important rivers, lakes and glaciers of Kumaun have been discussed in a separate chapter.

LANDFORMS

1. *Formed by Ice*

Landform in the tracts lying above the snowline have been carved by the action of glaciers and avalanches. Moving ice acts as a corroding and transporting agent. This results in a typical landform. These include hanging valleys, cirques, horns, cols, U-shaped valleys, glacial lakes, waterfalls, moraines (lateral, medial, terminal or recessional). Landforms carved by the action of ice are found in the upper valleys of the Dhauli, Kuti, Pindar, Gori and Kali rivers.

2. *Formed by Fluvial Action*

Landforms formed by fluvial action are found in tracts below the snowline. These include V-shaped valleys, gorges, canyons, rock benches, river cut and river built terraces and waterfalls. Such landforms have been formed by erosion, transportation and deposition by running water.

Rivers from the Himalayan slopes have deposited a large quantum of sediments in the bhabar and terai tracts and also in the dun valleys. Landforms carved out by fluvial action are seen in the middle and lower valleys of snowfed rivers and in the entire courses of the stream fed rivers of Kumaun.

3. *Ridges*

There are a number of prominent ridges in the lower and middle Himalayan of Kumaun. These include:

a) *Ranikhet Ridge*: This ridge towers above the Ramganga valley and follows its course for some distance.

b) *Almora Ridge*: The Almora ridge lies in south-central Kumaun. It towers above the surrounding mountains and valleys and includes Kausani.

c) *Nainital Ridge*: This ridge lies to the north of Nainital town.

d) *Bageshwar Ridge*: This is an almost north-south trending ridge that towers over the valley of the Sarju river.

SETTLEMENTS

About 83% of the total population of Kumaun lives in the rural areas while the rest lies in urban and semiurban settlements. The number of urban settlements in 1986 was 31 and that of villages was about 7000.

1. *Rural Settlements*

The 7000 rural settlements of Kumaun include inhabited forest settlements. The population of these settlements is very low in the high hills and high in the terai and bhabar tract. They may be placed into groups on basis of the size of population:

Population size	*% of total*
Less than 200 persons	55.0
200 to 499 persons	30.8
500 to 999 persons	10.5
1000 to 1999 persons	03.1
2000 to 4999 persons	00.6
Total	100.00

The sites on which rural settlements are common include: (a) river valleys with broad terraces, (b) transverse spurs, and (c) gently sloping plains of the terai and bhabar tract.

There are two types of rural settlements commonly seen in Kumaun. The agglomerated or nucleated settlements have one or more cluster of houses in the village. In the case of dispersed settlements there is dispersion of houses due to terrain or intervening forests. Under these conditions land-holdings are fragmented and dispersed.

In many parts of Kumaun, twin villages are also common. People own houses in both villages which are situated in the valleys and also on mountain tops. The former are used in winter and latter in the hot season.

2. *Urban Settlements*

In 1981 the number of urban settlements in Kumaun was 26. This rose to 31 in 1986 and is estimated to be 35 in 1990. The twin towns of Haldwani-Kathgodam have the largest population in Kumaun. The grouping of population size in the 31 urban settlements of Kumaun is given below:

Size of population (in persons)	*Number of settlements*
Less than 5000	15
5000 to 20000	09
20000 to 50000	05
50000 to 100000	02
Total	31

Some of the larger urban settlements of Kumaun are Haldwani-Kathgodam, Kashipur, Rudrapur, Tanakpur, Nainital, Almora, Rani-khet and Pithoragarh.

LAND USE

Kumaun has a total geographical area of 21,035 sq. kms. The districtwise distribution of land is:

District	*Land area (in sq. kms)*
Almora	5,385
Nainital	6,794
Pithoragarh	8,856

The districtwise land use is given below:

Almora District

Land use	*% of total area*
Forest area (recorded)	52.2
Pastures	06.9
Cultivated area	18.9
Uncultivable area	04.6
Cultivable wastelands	10.3
Orchards	05.2
Other uses	02.9

Pastures occupy more area in this district while the cultivated area is relatively less.

Nainital District

Land use	*% of total area*
Forest area (recorded)	57.4
Pastures	00.2
Cultivated area	29.3
Uncultivable area	00.9
Cultivable wastelands	05.7
Orchards	02.2
Other uses	04.3

Bulk of the cultivated area in this district is in the terai and bhabar tract.

Pithoragarh District

Land use	*% of total area*
Forest area (recorded)	50.4
Pastures	12.3
Cultivated area	13.5
Uncultivated area	03.8
Cultivable wastelands	10.9
Orchards	06.9
Other uses	02.2

Pastures cover a fairly large proportion of the land area in this district. This includes the high altitude pastures near the snowline.

Land Capability

Detailed land capability studies are lacking. The following classes are recognised (after Jalal 1988):

Land capability	*Area in hectares*
A. Good quality low land	6,86,651
1. Irrigated low land	4,91,993
2. Unirrigated low land	1,45,991
3. Water logged low land	48,667
B. Medium quality upland	1,61,828
4. Superior upland	97,505
5. Inferior upland	64,323
C. Poor quality hill land	1,36,691
6. Unterraced hill land	2,484

7. Shifting cultivation land	1,065
8.Pasture land	1,33,141
D. Forest land	11,27,362

Land capability has also been discussed in the chapter on Economy.

ADMINISTRATIVE UNITS

Kumaun is a civil division headed by a divisional commissioner whose headquarters are located at Nainital. Kumaun division is made up of the following districts:

1. Almora District

This district encompasses central Kumaun. Its headquarters are located at Almora, an old town situated atop a 1800 mts high ridge. It is well known for its rich forests and for picturesque hill stations which attract thousands of tourists each year. Kausani is famous for its snow views. Below it is the flat valley of the Gomati river while at Baijnath there are a number of ancient temples.

The garrison town of Ranikhet is also much frequented by visitors. Agriculture is the mainstay of the economy of Almora district. It produces wheat, maize, paddy, vegetables and potatoes. Apples and other fruits are also grown in the hills and valleys.

The development blocks of Almora district are:

1. Bageshwar
2. Bhainsiachhana
3. Chaukhutia
4. Dhauladevi
5. Garur
6. Hawalbagh
7. Takula
8. Bhikiasen
9. Dawarahat
10. Kapkot
11. Lamgara
12. Salt
13. Sialde
14. Tarikhet

Important towns of Almora district are Almora, Ranikhet, Bageshwar, Kausani and Baijnath.

2. Nainital District

This district occupies the south-eastern and southern part of Kumaun. Its headquarters are located at Nainital. This district is well known for beautiful lakes and hill stations that have developed around them. The southern tract of this district is made up of the well developed terai and bhabar tract which has made rapid strides in agriculture and industry.

There is a wide network of roads in Nainital district. Agriculture is the main source of livelihood for the people of this district. The principal crops grown in Nainital district are paddy, maize, sugarcane, wheat and pulses. Many industries have been set up in the terai and bhabar tract. Fruits are also grown in this district.

The development blocks of Nainital district are listed below:

1. Bazpur	2. Gadarpur
3. Haldwani	4. Kashipur
5. Ramnagar	6. Rudrapur
7. Bhimtal	8. Khatima
9. Betalghat	10. Kotabagh
11. Okhalkanda	12. Dhari
13. Ramgarh	14. Sitarganj

Important towns of this district are Nainital, Haldwani, Kathgodam, Bhowali, Ramnagar, Rudrapur, Kashipur, Khatima, Bazpur and Bhimtal.

3. Pithoragarh District

The Pithoragarh district occupies the northern part of Kumaun. Its borders touch the international border with Tibet in the north. The headquarters of this district are located at Pithoragarh. The northern part of Pithoragarh district is almost entirely covered with snow. There are a number of remote valleys such as the upper Darma and Gori Ganga. A large number of glaciers descend down from the main Himalayan range that forms its northern boundary.

The traditional road leading to the Mansarovar lake passes through Garbyang and Dharchula before ascending to the Lipu Leh pass and across it into Tibet. The southern part of this district is more developed and has a good network of roads. Agriculture and animal husbandry are the mainstay of the economy of Pithoragarh district. Amongst the crops grown are maize, barley, jowar, ginger and potatoes.

The development blocks of Pithoragarh district are listed below:

1. Binh	2. Berinag
3. Didihat	4. Gangolihat
5. Munakot	6. Kanalichhina

7. Munsiari
8. Barakot
9. Champawat
10. Lohaghat
11. Dharchula
12. Paati

Important urban and semi-urban centres of Pithoragarh district are Pithoragarh, Champawat, Berinag, Askot, Didihat, Dharchula, Tawaghat and Garbyang.

DEMOGRAPHY

Population Growth

There has been a steady increase in the population of Kumaun during the last 130 years or so. This has been brought out in the following table:

Year of census	Population (in crores)	% increase or decrease during the last 10 yrs
1865	0.03	–
1869	0.04	–
1901	0.07	–
1911	0.08	+ 9.2
1921	0.08	– 4.8
1931	0.09	+ 6.6
1941	0.10	+ 13.9
1951	0.11	+ 13.2
1961	0.15	+ 33.1
1971	0.18	+ 26.0
1981	0.24	+ 28.5

This table shows that there has been a steady increase in the population of Kumaun. The rate of growth went up sharply after 1951. In the terai and bhabar tract of Nainital district there has been an increase of nearly 300 per cent between 1951 and 1981. The district-wise rate of population growth between 1951 and 1981 is given below:

District	% increase from 1951 to 1981
Almora	56.0
Nainital	71.0
Pithoragarh	236.0

The excess of births over death has apparently contributed to this enormous growth which is more than the national average particularly in case of Nainital and Pithoragarh districts.

Migration

A part of the population growth in the middle and upper hills was offset by the migration of people away from this tract for various reasons. The trends of migration in different districts is given below (after Khanka, 1988):

District	Year	In-migration (percentage)	Out-migration (percentage)	Net migration
Almora	1951	1.9	3.0	– 1.1
	1961	3.3	16.8	– 13.5
	1971	3.6	13.0	– 9.4
	1981	5.0	14.1	– 8.1
Nainital	1951	39.6	NA	–
	1961	33.1	8.0	+ 25.1
	1971	24.3	4.8	+ 19.5
	1981	31.7	5.2	+ 26.5
Pithoragarh	1951	–	–	–
	1961	3.1	5.4	– 2.3
	1971	4.0	10.6	– 6.6
	1981	5.6	11.6	– 6.0

Population Density

The population density in Kumaun depends on the following variables:

—natural factors like climate, topography, soil fertility

—level of development, industrial productivity, state of agriculture

—availability of facilities like health and education

—employment opportunities available.

The population density of different districts of Kumaun (1981)

District	Population density (persons per sq. km)
Almora	
total	141
rural	133
urban	1262
Nainital	
total	167
rural	123
urban	3769
Pithoragarh	
total	55
rural	52
urban	757

Age Structure

In Kumaun the age structure of the population is broad based. There are a high proportion of people in the younger age groups. This is due to a high birth rate and a low, steadily declining rate of infant mortality. The age structure of the population in different districts of Kumaun is given in the following table:

District	Age group (in years)	Population (%)
Almora	0-14	40.87
	15-34	28.91
	35-59	22.60
	above 60	07.62
Nainital	0-14	41.91
	15-34	32.66
	35-59	19.92
	above 60	05.51
Pithoragarh	0-14	39.68
	15-34	30.79
	35-59	22.44
	above 60	07.90
Kumaun division	0-14	41.12
	15-34	31.90
	34-59	21.29
	above 60	06.50

Urbanization

This pertains to the change in the population living in the urban areas. In Kumaun, like most other parts of India, there has been an increase in the population living in the urban areas. This has been amply brought out in the table given below:

District	1951 (in per cent)	1981	% increase (from 1961)
Almora	4.85	6.29	102.25
Nainital	22.36	27.49	316.60
Pithoragarh	–	5.52	126.25

2
Climate

Climatic conditions of Kumaun vary from tropical in the bhabar tract to arctic on the higher Himalaya and the inner dry upper Darma valley. While the lower hills, dun valley, Siwalik hills and terai and bhabar tract receive very heavy rainfall from the SW monsoons, the upper Darma valley is virtually devoid of rain except for a few showers.

Till a few decades back, whatever knowldege we had of the climatic conditions prevailing in Kumaun was based on the scanty data of the forest department which had meteorological instruments at most range headquarters. Detailed records were available only for hill stations frequented by the British, e.g. Nainital and Ranikhet.

However, today a fairly detailed level of knowledge is available for the climatic conditions prevailing in different parts of Kumaun. Weather satellites beam back pictures of the cloud cover over this region or indicate the snow cover and even the movement of glaciers on the upper Himalayan slopes.

The following monsoon systems of the Asian continental mass has a profound impact on the weather and climate of Kumaun and the adjoining hills.

1. *SW Monsoons*

There are gigantic land and sea breezes that blow across the Indian sub-continent with clock work regularity once every year. The SW monsoon winds originate due to the differential behaviour of the Asian landmass and the sea to the incoming solar radiation. In summer, the heartland of the Indian sub-continent becomes very hot as the mercury soars to over 40°C in central and western India and most parts of Pakistan except in the north and north-west. During this season there is a low pressure over this tract and a high pressure over the mass of water surrounding the sub-continent, viz. the Arabian sea, Indian Ocean and Bay of Bengal. As a result of this phenomena, winds from the high pressure over the water mass blow in towards the low pressure over the land mass. While passing over the seas and oceans, these winds take up moisture and become saturated with it. These moisture laden clouds are known as the SW monsoons and they cause heavy and widespread rains in most parts of Kumaun from the middle of June to September.

N.E. Monsoons

This cycle is reversed in winter when there is a high pressure over Central Asia with a corresponding low pressure over the oceans and seas in the south. This results in winds blowing in the opposite direction. These winds are known as NE monsoons and they cause winter rains in many parts of Kumaun. However, these rains are not as heavy as are experienced in other parts of the western Himalaya as the NE monsoons are considerably weak in this part of the Himalayan mountain chain.

Cycle of Seasons

Like other parts of the Himalaya, Kumaun too experiences a well defined climatic or seasonal cycle.

1. *Winter*

The winter season is perhaps the longest season in Kumaun. It begins in mid-November and lasts till the end of March. This is also the severest season in this region as temperature drops down to around the freezing point, in many parts of this tract. It remains below 0°C in parts of the higher Himalaya and the inner dry upper Darma valley. Snowfall commonly occurs in tracts lying above an elevation of about 2200 mts. Frost is experienced in the valleys and terai and bhabar tracts.

In winter, the northern cold dry winds and western disturbances also have a bearing on the climate of Kumaun. There occur avalanches and snowstorms in the higher reaches near the line of perpetual snow. People living at high altitudes are forced to move down to the shelter of the valleys in winter as these tracts are covered by a thick blanket of snow for long periods.

The temperature usually begins to get warmer in the middle of March.

2. *Spring*

The spring season marks the transition from winter to summer. It extends from the middle of March to late April or early May. This season is very short in the upper tracts of Kumaun where the winter season may extend till early May and summer directly sets in with virtually no transition period in between.

There occur local rains and windstorms in this season. Frost is experienced in the early part of the spring season. In spring, tracts occurring above an elevation of about 3000 mts may experience snowfall. The nights are usually clear in spring.

3. *Summer*

This is one of the three main climatic seasons for Kumaun. It extends from late April or early May to late June, when the monsoon rains set in. It becomes very hot in southern Kumaun, viz. the terai and bhabar tracts, dun valleys, Siwalik hills and the foothills, during the summer season. The mean maximum temperature may be more than 40°C.

However, tracts lying above an elevation of about 3000 mts experience a relatively mild summer. Days are warm and nights are cooler and pleasant. The mean maximum temperature is usually not more than 28°C. There may also occur occasional showers in the summer season.

In southern Kumaun, dust from the adjoining Ganga plains rises up and covers the sky, at times reducing visibility to barely a few metres. There are conventional and orographic stormy disturbances accompanied by thunder and lightning during the beginning and end of this season. The hot dusty *loo* winds also effect the lower hills of Kumaun. They have a desiccating effect.

Pre-monsoon showers in mid-June mark the arrival of the SW monsoons.

4. *Monsoon*

Together with winter and spring, this is the third main climatic season experienced in Kumaun. The south-west monsoons cause heavy and widespread rains in most parts of Kumaun from late June to the middle of September. The monsoon arrives earliest in eastern Kumaun. There may be a difference of about 5 days in the arrival of the SW monsoons in eastern and western Kumaun. The SW monsoons account for bulk of the total annual precipitation in this hilly region.

Very heavy rains are received in the south facing slopes during the monsoon season. In the valleys, rains may continue for several days at a stretch. Mountainsides are covered for long periods by a thick blanket of mist during which time the sun rarely comes out. Flash floods and landslides are common in the monsoon season. As a result there is widespread damage to life and property. Many far-flung tracts remain cut off from the rest of the world during the season.

The total rain caused by the SW monsoons is more than 200 cms in most parts of Kumaun.The force of the monsoons weakens in September and the sun comes out more frequently. The effect of the SW monsoon finally ends by the middle or late September.

Even as most parts of Kumaun are experiencing very heavy rainfall, there are some areas where the effect of the SW monsoons is negligible. These are the rain shadow areas of the main Himalayan range, e.g. the upper Darma valley. They remain devoid of rains in the monsoon season.

5. *Autumn*

The autumn season marks the transition between the monsoon and winter seasons. It extends from the middle of September to the beginning of the winter season. This is the best season of Kumaun as the weather clears up after the long rainy season. Both the day and night are clear. The weather remains exceptionally fair in most parts of Kumaun except at very high elevations where the first snowfall of the season may take place in early October.

There is virtually no frost in autumn. The days are warm and nights are cool and pleasant. Little rainfall may be experienced in some parts of Kumaun towards the end of the rainy season.

Climatic Regions

Based on altitude, latitude and prevailing climatic conditions, Kumaun may be divided into the following climatic regions:

Region	*Altitudinal range (in mts)*
Tropical	below 500
Sub-tropical	500 to 1500
Sub-temperate	1500 to 2200
Temperate	2200 to 3500
Sub-arctic	3500 to 4500
Arctic	above 4500.

Factors Affecting Climate

The following factors have a bearing on the climatic conditions prevailing in different parts of Kumaun.

1. *Altitude or Elevation*

This is one of the most important parameters influencing the prevailing climatic conditions. It influences:

—the amount, distribution and intensity of rainfall
—wind speed and direction
—snowfall and its frequency
—frost and mist
—relative humidity and its fluctuation.

There is a change in the temperature after every 200 mts variation in elevation. This change is more faster at higher elevations. In the same way the total annual rainfall received increases with elevation upto about 2500 mts above which snow accounts for significant portion.

On the other hand, there is very low rain in tracts lying in the rain shadow of the main Himalaya even though they have a high elevation.

One remarkable feature of high altitudes is that the ground exposed to sunlight is heated very rapidly and intensely and on the other hand the shaded area remains severely cold.

2. *Aspect*

The influence of aspect on the climatic conditions of Kumaun has been brought out in the following points:

a) More sunlight and rain is received by the south facing slopes.

b) Even in the lower and middle Himalaya, the slopes lying behind high ridges are shady and fall in the rain shadow zone.

c) South facing slopes receive less snowfall than their northern counterparts even at the same elevation.

d) In the higher Himalaya, the snowline is at higher elevation on south facing slopes than on the northern slopes of the same mountain.

e) Climatic conditions are by and large warmer on southern aspects.

3. *Local Relief*

Local relief too has an influence on the climatic conditions prevailing in an area or tract.

a) A slope facing towards east experiences warm mornings and cool afternoons while west facing slopes experience cool mornings and warm afternoons.

b) Inversion in temperatures are observed when air cooled by radiation at night flows downhill into the valley bottom and settles there. Thus, undisturbed by winds, this cool air reduces the temperature of the valley bottom at night by several degrees as compared to nearby tracts on the top of the ridge.

c) The morning sun dissolves this mist and forces it to rise upwards in the form of cumulus or stratocumulus clouds.

d) The diurnal range of temperature is more at the valley bottom and less at mountain tops.

e) There is more frost in the valley bottom as compared to mountaintops.

Peculiar Climatic Phenomena

Like other parts of the western Himalaya, the Kumaun hills too experience peculiar or special climatic phenomena. These have been discussed in the following text:

1. *Gravity Winds*

Gravity winds are vertical movements of air or winds that are generated due to changes in temperature conditions during different parts of the day. These winds are an important part of the variation in climatic conditions prevailing in different tracts of Kumaun. The salient features of this climatic phenomena are:

a) There is a downward movement of air from the mountain top to the valley bottom after sunset every day.

b) At night, clouds and mist settle down in the valley bottom and remain there till day break the next day. Thus, the skies become clear and starry at the top of the ridge.

c) At day break the morning sun warms the air as a result of which the mist dissolves and the warm air rises up in the form of giant clouds.

d) As the day advances, high energy anabetic winds start blowing up the mountain slopes. They attain very high speeds at elevations of over 6000 mts. They may blow fine particles of snow with them.

e) The direction of wind is reversed towards evening and high energy katabatic winds blow down the mountain slopes. They descend in the form of furious gusts of cool air particularly along gaps in the mountains formed by rivers and streams.

2. *Western Distrubances*

The global westerly winds of the mid-latitudes also make their presence felt in parts of Kumaun. These disturbances have a bearing on precipitation in the winter season in the following manner:

a) They cause rain at lower elevations and snowfall in the upper reaches.

b) Blizzards and snow storms are caused at very high elevations in Kumaun.

c) Warm winds blowing in from the south-east mark the approach of the western disturbances. Later on there occur cirrus and cirro-cumulus clouds, thereby filling the grey skies with stratus clouds.

d) There may be drizzling rain at lower altitude; sleet at mid-altitude and snowfall and blizzards at higher elevations.

e) In Kumaun, the western disturbances account for 10 to 12 per cent of the total annual precipitation.

3. *Visibility*

In May and June, there occurs a haze of dust in the lower hills. Even though there are virtually no clouds in the sky, visibility is very poor as there is a haze of dust which may extend upto an elevation of 3000 mts. The atmospheric turbidity at this elevation is very high. This dust is blown into the skies of Kumaun by the hot dusty winds of the Ganga plains.

On some days the atmospheric turbidity occurring in lower Kumaun is so high that it is slightly less than that prevailing in adjoining parts of the Ganga plains.

4. *Jet Streams*

As the name suggests, these are winds blowing at very high speed which develop in a narrow belt of the upper atmosphere in winter, at very high elevations. These winds may attain speed of over 200 kms per hour. They increase in speed and force with height above the earlier surface upto about 10 kms above which the speed and force decreases.

5. *Lee Waves*

These air currents on the leeward side of a mountain. They attain very high speed and are capable of causing considerable erosion and damage.

Factors influencing winds: Winds are an important climatic phenomena in Kumaun. They vary from the SW monsoon winds to the jet streams and leewaves. The factors listed below influence the speed, force and even direction of winds in Kumaun:

—speed profile of the wind
—stability of the atmosphere
—prevailing climatic conditions
—altitude or elevation at which the winds are blowing
—shape and size of the obstacle, if any
—stratification of the air mass.

Climate of Some Stations

The climatic conditions of some stations in Kumaun have been given in the following text:

1. *Kicha*

Kicha experiences a typical tropical climate. The mean maximum temperature is more than 40°C. December and January are the coldest months with a mean monthly temperature of 21°C. The total annual rainfall received in Kicha is about 140 cms.

2. *Ramnagar*

Ramnagar experiences climatic conditions varying from tropical to sub-tropical. The mean monthly temperature in summer is 39°C while that in winter is 20°C. The total annual rainfall is 148 cms.

3. *Tanakpur*

Tanakpur too experiences a tropical to sub-tropical climate. The mean monthly temperature in summer is 40°C while that in winter is 22°C. The total annual precipitation is about 205 cms.

4. *Mukteshwar*

A typical temperate climate prevails in this town. The mean monthly temperature in summer is 19°C while that in winter is 6°C. The total annual precipitation is about 135 cms.

5. *Nainital*

Temperate climatic conditions prevail in Nainital. The vast water body has a moderating effect on summer temperatures and the monthly mean temperature is 20°C. In winter, the monthly mean temperature is 6°C. The total annual precipitation received in Nainital is about 250 cms.

6. *Ranikhet*

This town too experiences temperate climatic conditions. The mean monthly temperature in summer is 22°C while that in winter is 8°C. Ranikhet receives about 135 cms of precipitation each year.

7. *Almora*

Almora also experiences a temperate climate. The mean monthly temperature in the summer season is about 24°C while that in winter is 8°C. The total annual precipitation recieved in Almora is about 110 cms.

8. *Pithoragarh*

Temperate to sub-arctic climatic conditions prevail in Pithoragarh. The mean monthly temperature ranges from 23.5°C in summer to 7.5°C in winter. The total annual precipitation received in about 125 cms.

9. *Dharchula*

The climatic conditions prevailing in Dharchula vary from temperate to sub-arctic with the mean monthly temperature ranging from about 24°C in summer to 8°C in winter. The total annual precipitation is 150 cms.

10. *Munsiari*

Temperate to sub-arctic climatic conditions influence Munsiari. The mean monthly temperature in summer is 22°C and that in winter is 6°C. The total annual precipitation is about 215 cms in Munsiari.

11. *Haldwani-Kathgodam*

The twin towns of Haldwani and Kathgodam experience climatic conditions varying from tropical to subtropical. Summers are very hot and the temperature may soar to over 40°c. In winter the mercury drops down to about 3°c or even lower. Frost is common in the winter season. The total annual precipitation is about 150 cms.

12. *Kausani*

Kausani experiences a typical temperate climate. The mean monthly temperature in summer is 20°c while that in winter it is 5°c. Very heavy snowfall is received during the winter season. The total annual precipitation received at Kausani is about 160 cms.

13. *Bhowali*

A typical sub-tropical climate is experienced in Bhowali. The mean monthly temperature in summer is 30°c which drops down to about 6°c in winter. The total annual precipitation is about 180 cms.

3

Soils

Soil is the base on which all animal and plant life survives. In fact the life and well being of the human civilization is dependent either directly or indirectly on soil. The soils of Kumaun vary from the rich and deep alluvial soils of the terai and bhabar tract to the thin and bare soils of the high mountains and almost desert like soils of the trans-Himalayan and inner dry valleys.

The nature, composition and distribution of soils in Kumaun depends on the following parameters.

1. The nature and composition of the underlying country or parent rock.
2. The process by which that particular soil has formed.
3. Prevailing climatic conditions particularly the distribution and intensity of rainfall.
4. The nature and composition of the vegetation supported by the soil.

Soils in turn have a profound impact on the life of the people of Kumaun. This may be summarised in the following points:

—it provides a base for growing food
—forests and wild animals survive on the soil
—soils also play an important role in providing shelter to human beings.

The principal soil types found in different parts of Kumaun have been discussed in the following text (after Ray Chaudhri, 1960 and Negi, 1990):

1. Himalayan Alluvial Soil (Group A)

These are alluvial, riverine, non-calcerous to moderately calcerous soils found in the following parts of Kumaun:

1. longitudinal dun valleys
2. terai and bhabar tract
3. valleys of the rivers in the lower and middle Himalaya.

These soils are red in colour, deficient in lime, humus and phosphoric acid. The topsoil is lacking in phosphate and calcium and has a varied nitrogen content. The texture ranges from clayey to sandy loam while the reaction of this soil is largely acidic.

Such soils are fairly old in origin. However they are limited in extent as in many localities these soils have been overlain by more recent soils. The principal characters of a typical section of this soil in the high rainfall terai and bhabar tract of Nainital district is given below:

Horizon	Soil depth (in cms)	Characters
A	10 to 20	light yellowish brown, loamy, angular, blocky, plastic, plant roots are abundant.
B	20 to 25	yellowish brown, clayey, angular blocky, sticky, friable, plant roots abundant, hard when dry.
C	25 to 40	dark yellowish brown, angular, blocky, very plastic and sticky, mildly alkaline.
D	40 to 65	yellowish brown, clayey, angular, blocky, mildly alkaline.

2. Himalayan Alluvial Soil (Group B)

These are alluvial, riverine soils of recent origin. They are found in the following parts of Kumaun:

1. terai and bhabar tract
2. dun valleys
3. along streams, rivers and on river built terraces.
4. uplifted river terraces.

The Himalayan alluvial soil (group B) are lacking in nitrogen, humus and phosphoric acid. The top soil is less alkaline in reaction than the layers lying below. These soils are of recent origin and may overlie the older soils.

The main characters of a typical soil profile in Nainital district is given below:

Horizon	Soil depth (in cms)	Characters
A_1	0 to 30	olive green, fine sandy loam, single grained, loose, soft, non-sticky.
A_2	30 to 40	olive greyish green, loamy sand, single grained, calcerous, moderately alkaline
B	40 to 55	olive grey, loamy sand, calcerous, moderately alkaline
B_1	55 to 75	olive grey, fine sandy texture, non-sticky, moderately alkaline.

3. Red and Black Soils

These soils have a limited distribution in Kumaun. Red soils are restricted in a few pockets, mainly over acid igneous rocky, viz. granite. Such soils are developed over the Amritpur and Nainital granites.

Red and black soils may further be of two sub-types:

1. Red soil dominant—light textured, devoid of carbonates
2. Black soil dominant—less clayey, defficient in lime.

These soils are lacking in phosphate, nitrogen, humus and lime. Their depth ranges from 1 cm to 75 cms. They support conifer and broadleaved forests and scurblands. The main characters of various horizons of a typical red and blacksoil of a high rainfall tract in Almora district is given below:

Horizon	Soil depth (in cms)	Characters
A	0 to 15	reddish brown, sub-angular blocky, sandy gray loam, hard and dry, permeable.
B	15 to 25	reddish brown, clayey loam, breaks into angular to sub-angular blocks, moderately permeable.
B_1	25 to 35	dark reddish brown, firm and moist, moderately permeable.
B_2	35 to 45	deep brown, clayey, low permeability.

4. Ferrugenous Red Soil

These soils are found in central Kumaun. They are well developed over the main Himalayan rocks, viz. quartzites, biotite schist, amphibolite schist etc. The red ferrugenous soil is free of carbonates, deficient in nitrogen, humus and phosphorous, light textured, porous, friable and have a very low proportion of friable soils. The soil depth ranges from about 10 cms to 75 cms.

These soils may be grouped into two on basis of morphology:

1. *Red earths*—having a loose, friable topsoil which is rich in secondary concretions.
2. *Red loam*—agrillaceous soils having a blocky structure.

Rich forests, scrub lands and pastures are found on these soils. The main characters of various horizons of a typical ferrugenous soil in Pithoragarh district is given below:

Horizon	Soil depth (in cms)	Characters
A	0 to 20	reddish yellow, loamy sand, friable, dry, roots of plants are abundant.
B	20 to 30	yellowish red, sandy loam, crumby structure, there may occur ferrugenous concretions.
B_1	30 to 40	yellowish red, sandy loam, compact, crumby.
B_2	40 to 60	yellowish red, sandy loam, blocky to crumby.

5. Brown Soil

These soils are common in many parts of Kumaun particularly under dense broadleaved temperate and sub-alpine forests. There occurs a thick layer of humus on the forest floor. It is made up of decomposed leaves, branches, twigs and even fallen twigs. The top soil too is extremely rich in humus.

The main characters of a typical soil profile of the brown soil is given below:

Horizon	Soil depth (in cms)	Characters
A	0 to 20	dark grey, extremely rich in organic matter, fine granular structure, colour becomes darker when wet.
B_1	30 to 45	brown micaceous, clayey, colour darkens when wet, organic matter content is relatively lower.
B_2	45 to 75	brown, micaceous, clayey, orgainc matter content is low.
C	75 to 90	brown, loamy.

6. Forest Soil

As the name suggests, this soil is found under conifer and deciduous forests of Kumaun. They are well distributed in the terai and bhabar tract, dun valleys, lower, middle and main Himalaya. Their formation has to a considerable extent been affected by the accumulation of organic matter in the top soil. This is in the form of leaves, twigs, branches and trees that fall on the forest floor.

Two conditions leading to the formations of forest soils are:

1. Mildly acidic or almost neutral conditions, high base status.
2. Acidic conditions, acidic humus and a low base status.

Different sub-types of the forest soil found in Kumaun are:

1. *Brown Forest Soil (Acidic to Neutral) under Pasturage*

The main characters of such soils are:

Horizon	Soil depth (in cms)	Characters
A	0 to 10	light yellow, loamy when dry, plant roots abundant.
A_1	10 to 20	brown, darkens when wet, plant roots relatively less.
B_1	20 to 35	yellowish brown, clayey loam, angular to sub-angular, blocky.
B_2	35 to 50	dark yellowish brown, angular to sub-angular blocky structure, clayey loam.
B_2C	50 to 70	brown, becomes dark when wet, gravelly loam, massive no mottlings and carbonates, fairly rapid permeability.

2. *Brown Forest Soil (Acidic to Neutral Virgin)*

Horizon	Soil depth (in cms)	Characters
A	0 to 15	dark brown, becomes very dark when wet, silty loam, fine granular tends to become friable when moist, roots abundant, rich in humus.
B	15 to 25	light yellowish brown, silty loam weak to moderate granular, loose when dry.
B_2	25 to 40	light yellowish brown, silty loam no well defined texture.
B_2C	40 to 60	yellowish brown, silty clay loam, permeability from moderate to fair.

3. Brown Forest Soil (Acidic) under Pasturage

Horizon	Soil depth (in cms)	Characters
A	0 to 10	brownish grey, loamy, very rich in humus, roots abundant.
B	10 to 20	reddish brown, loamy, humus content relatively less.
B_2	20 to 35	yellowish brown, sandy loam, humus absent.
B_3	35 to 50	yellowish grey, heavy texture, humus absent.
C	50 to 60	ash grey, loamy, moderate permeability, country rocks in lower layers of this horizon.

4. Brown Forest Soil (Acidic) Virgin

Horizon	Soil depth (in cms)	Characters
A	0 to 10	greyish brown, boundary with the lower horizon not very sharp, well defined crumby structure, no mottlings and concretions.
A_1	10 to 25	yellowish brown, distinct crumby structure, silty loam, no concretions and mottlings, permeability from moderate to fair.
B_1	25 to 40	dark brown, angular to sub-angular, clayey loam, permeability moderate.
B_2	40 to 60	dark brown, sub-angular blocky, clayey loam, no concretions.

7. Podsolic Soil

Podsolic soils have developed in humid temperate conditions usually under coniferous forests, e.g. deodar, blue pine, fir and spruce. These soils are found under the following conditions: (a) in the temperate, sub-alpine and alpine areas, and (b) over quartzites, granites, schists and gneiss.

The A horizon is distinctly leached where there is a deposition of organic matter in the B horizon. The soil depth varies from a few centimetres on steep slopes to about a metre. Iron podsols are more common in Kumaun.

The main characters of a typical podsolic soil in Almora district is given below:

Horizon	Soil depth (in cms)	Characters
A_0	0 to 10	black, sandy to clayey loam, rich in undercomposed organic matter, acidic reaction.
A_1	10 to 30	black, sandy to clayey loam, acidic in reaction.
A_2	30 to 50	deep black, sandy to clayey loam, acidic reaction.
B_1	50 to 70	brownish, sandy to clayey loam relatively more compact.
B_2	70 to 95	brown, sandy loam, large pockets of parent material occur in this horizon.

8. Foothill and Terai Soil

As the name suggests, this soil is found in the southern tract of Kumaun, viz.

—the terai and bhabar tract
—the dun valleys
—the foothills of the lower Himalaya.

These soils are acidic, poor in plant nutrients, deficient in phosphate. The colour varies from deep black to greyish black. Two chronic factors affect these soils: (a) heavy and frequent floods which lead to an increase in the proportion of micaceous sandy material, and (b) frequent spells of prolonged drought.

The principal characters of a typical section of this soil in Nainital district is given below:

Horizon	Soil depth (in cms)	Characters
A	0 to 15	dark grey, clayey loam, very rich in humus, moderately fine to crumby structure.
B	15 to 25	light greenish grey, clayey loam to silty loam, moderately fine crumby structure which becomes sticky and plastic when wet.
B_1	25 to 35	light yellowish, silty loam, very fine granular, neutral.
B_2	35 to 55	light yellowish, silty loam, very fine granular, becomes sticky and plastic when wet.

9. Mountain and Hill Soils

This is a collective name given to various types of soils found under the following conditions:

—at very high elevations
—under sub-tropical, temperate and sub-alpine conditions
—under varied forest types.

Mountain and hill soils are very thin, fertile and may be less than a centimetre deep on steep slopes. They are mixed with pebbles, shingles and gravels in many localities. Angular and sub-angular fragments of the parent rock may be found mixed with the lower layers of the mountain and hill soils. Texture varies from loamy to sandy loam. The soil reaction is from acidic (pH 4.6 to 6.5) to neutral. The soil organic matter content varies from 1 to 5 per cent.

The main characters of a typical profile of the mountain and hill soil in a part of Almora district is given below:

Horizon	Soil depth (in cms)	Characters
A	0 to 20	light brownish grey, sandy loam, granular, friable, no mottlings, moderate permeability.
A_1	20 to 65	yellowish brown, moist, no mottlings, clay loam, blocky, very hard and compact, permeability low.
B_1	65 to 95	yellowish brown, moist, clay loam, no mottlings (except occasional grey and brown mottlings), extremely firm and difficult to plough, roots more or less absent, low permeability. Fragments of the parent material may be found mixed with the lower layers of this horizon.
B_2C	95 to 105	yellowish brown, no mottlings except in freshly cut surfaces where brown and grey mottlings are noted, clayey loam, laminated and compact free of carbonates, poor permeability, fragments of parent rock abundant.
C	over 105	yellowish brown, the lower layers of this horizon are underlain by the parent rock, permeability very low.

10. High Altitude Meadow Soil

The high altitude meadow soils are usually found near the snow-line in all parts of northern Kumaun, viz. the upper tracts of Almora and Pithoragarh districts. They are found both in the higher Himalaya and trans-Himalaya. In both these tracts, bulk of the total annual precipitation is in the form of snow which may lie on the ground for several months at a stretch in the cold season.

The high altitude meadow soils are found under two types of vegetation:

—under high altitude pastures or moist alpine scrub
—under dry alpine scrub or xerophytic vegetation of the trans-Himalayan tract.

These soils are very thin and fragile. They are rocky in composition and are highly prone to displacement as a result of slides, creeps and avalanches. Soils under alpine meadows are very dark in colour and rich in humus. Permeability ranges from rapid to moderate.

4

Rivers, Lakes and Glaciers

Kumaun is extremely rich in water resources, viz. rivers, streams, lakes and glaciers. This water is used for drinking, washing, irrigating and generating power not only in these hills but also in the plains of northern India.

RIVERS

Two main river systems drain Kumaun. These are Kali and Ramganga river systems. They are a part of the Ganga drainage basin. Thus, the waters from the Kumaun hills flow into the Bay of Bengal.

Rivers of Kumaun originate in the snows on the slopes of the main Himalayan range in the north. Some rivers and streams also originate as springs fed by underground water.

Kali River System

The river Kali or Kaliganga drains eastern Kumaun and western Nepal. It flows for a considerable distance along the border of Kumaun (India) and Nepal. This river is formed after the confluence of two main headwaters:

—the river Kalapani is its eastern headwaters. It is made up of numerous small springs fed by underground waters.

—the river Kuthi Yankti forms the western headwaters of the river Kali. It rises in the snowfields of the Himadri on the southern slopes of the great or main Himalayan range.

Both these headwaters join to form the Kali at the base of the main Himalayan range. Thereafter, the river flows in a SSW direction along a deep V-shaped valley to enter the plains of northern India at Baramdeo. It flows into the Ghagra in the plains of Bihar where this river is known as the Sarda.

The valley of the Kali river is steep. There occur river terraces of various sizes along it. Important tributaries of this river draining Kumaun are:

—the river Dhauliganga draining north-eastern Kumaun
—the river Goriganga draining north-central Kumaun
—the river Sarju draining central Kumaun
—the river Ladhiya draining south-central Kumaun.

Small towns and large villages that have come up along this river are Lipulekh, Garbyang, Dharchula, Jauljibi and Tanakpur.

1. *Kalapani River*

This river is the eastern headwaters of the Kali river near the Nepal-Kumaun border. It is made up of numerous smaller streams.

2. *Kuthi Yankti River*

This is the western headwaters of the Kali river near the Nepal-Kumaun border. It rises in the snowfields of the Himadri on the southern slopes of the great or main Himalayan range.

3. *Dhauliganga River*

The Dhauliganga river is a tributary of the Kali river in Pithoragarh district of Kumaun. It rises as two snow-fed mainstreams from valley glaciers on the southern slopes of the Kumaun-Tibet water divide. These two channels join after emerging from their U-shaped valleys. Thereafter this river flows in a general direction towards south-east to merge with the Kali river flowing in from the north near Dharchula.

The upper catchment of this river has been carved by the action of glaciers. This is a rain-deficient tract. Small hanging glaciers containing side glaciers open into the main valley at many places. The snow-melt waters from these glaciers join the Dhauliganga river at various points in its upper and middle courses.

The Dhauliganga river flows along a deep, V-shaped valley in its middle and lower courses. There occur small river terraces along this river. These are under cultivation. Alpine and sub-alpine forests cover the upper catchment of this river.

4. *Darma River*

This is a snow-fed tributary of the Kali river system in northern Kumaun. Its valley lies in the rain shadow of the main Himalayan range and hence is a dry tract receiving very low rainfall. The valley of this river is U-shaped.

5. *Goriganga River*

The Goriganga river is an important tributary of the Kali river draining northern Kumaun. It rises from near Unta Dhura on the Alaknanda-Kali water divide. This is a snow-fed river whose upper catchment is glaciated. It flows along a south-easterly course and joins the Kali river near Jauljibi.

The Goriganga river flows along a U-shaped valley in its middle and lower courses. It is joined by a number of snow and rainfed tributaries coming in from the north-east and south-west. The gradient of the upper course of this river is very steep. It becomes gentler near the mouth of the river.

River terraces formed by deposits brought down by running water are found on either banks of this river. Other prominent geomorphic features occurring in its middle and lower catchment include incised meanders, gorges, rock benches, steep cliffs and waterfalls.

The vegetation occurring in the catchment of this river vary from alpine meadows below Unta Dhura to sub-alpine, temperate and sub-tropical coniferous and deciduous forests.

6. *Sarju River*

The Sarju river is the largest tributary of the Kali river draining Kumaun. It rises in the area to the north-west of Baijnath in central Kumaun. This river flows towards south in its upper course, then turns towards south-east and joins the Kali river near Pancheshwar.

This river and its network of tributaries is fed by springs emanating from reservoirs of underground water. The entire valley of the Sarju has been shaped by the action of running water. The prominent geomorphic features include incised meanders, terraces and rock benches.

The Ramganga (Sarju) river is an important tributary of the Sarju river. It flows in from the north and joins the Sarju river in its lower course to the north-west of Pancheshwar.

The Sarju river flows along the northern base of the Almora massif. The entire catchment is covered by temperate and sub-tropical forests that are under severe biotic pressure due to the high population density in this tract. A number of large human settlements have come up along this river. Amongst these are the temple towns of Baijnath and Pancheshwar.

7. *Ramganga (Sarju) River*

This river is a tributary of the Sarju river. It rises from a small glacier on the south-eastern slopes of the Garhwal-Kumaun water divide. Thereafter it flows in a general direction towards south-east before flowing into the Sarju river, a short distance upstream of the mountain hamlet of Pancheshwar.

The upper catchment of this river has been shaped by the action of glaciers. Moraines occur on the valley bottom. These have been deposited by glacial action in the past. The valley becomes narrower once the river enters its middle course as this tract has been almost entirely carved by the action of running water. The river bed is strewn with boulders of various sizes. River terraces occur all along its course.

The Ramganga (Sarju) river has carved a V-shaped valley in its middle and lower courses. Alpine, sub-alpine, temperate and sub-tropical forests cover its catchment. Many large villages have developed along its course.

8. *Ladhiya River*

The Ladhiya river is a small tributary of the Kali river. It rises as two small springs fed by underground waters in the area to the north-north-east of Nainital. This river flows towards south-east in its upper and middle courses, turns towards east and flows into the Kali river upstream of Tanakpur.

Two small springs are the headwaters of this river. They rise at a distance of about 20 kms from each other. Thereafter they merge and flow towards south-east. A number of smaller streams flow into the Ladhiya river, particularly from the northern flank of its valley.

The valley of the Ladhiya river is deep and V-shaped in its upper course but opens up towards its mouth. Many small paired and un-paired terraces are found along the valley of this river. These are made up of deposits laid down by the river over the past thousands of years. Broad terraces are found near its confluence with the Kali river.

Forests of varying types and compositions occur in the catchment of this river. These include both deciduous and coniferous forests.

Large villages are located along the Ladhiya river.

Ramganga River System

This river system drains south-western Kumaun. Its catchment area in Kumaun is relatively small. The Ramganga river is a part of the Ganga system. The main streams of this river rise from the middle Himalayan slopes on the south eastern flank of the water-divide with the Alaknanda river system. It is essentially fed by springs emanating from reservoirs of sub-surface waters. Channels from a number of such springs merge to form the mainstream of this river.

Thereafter it winds its way across the lower Himalayan hills of Almora district. This river has carved a deep V-shaped valley. Amongst the prominent geomorphic features found in this tract are incised meanders, paired and unpaired terraces, interlocking spurs, waterfalls, rock benches, cliffs and towering ridges.

Forests are found in the entire catchment of this river, particularly as it flows through the dun-type valley of the Corbett National Park. Large settlements along the Ramganga river are Ganai and Dhikala.

1. *Kosi River*

The Koṣi river is an important river of the foothills of Kumaun. It rises as two springs fed channels in the lower Himalaya near Almora. The two channels merge after a short distance and the Kosi river flows towards south-west before making a U-trun and flowing in a south-easterly direction and entering the plains of western Uttar Pradesh downstream of Ramnagar.

A number of small tributaries flow into the Kosi river at various places from its source to where it enters the plains. This river has carved a V-shaped valley. Other geomorphic features of the Kosi catchment are incised meanders, interlocking spurs, river terraces at various levels, ridges, rock benches, cliffs and gorges. Small islands have formed at the confluence of the Kosi with its tributaries. It has cut deep gorges through the lower Himalaya and Siwalik hills.

Temperate and sub-tropical coniferous and deciduous forests cover the catchment of the Kosi river. Khair and sisham forests are found along the river bed. Small towns that have come up along the Kosi river are Garjia and Ramnagar.

2. *Gola River*

This is a small spring-fed river of the foothills of Kumaun. It drains the Siwalik hills to the east of Haldwani. This river rises as a small spring in the middle Himalaya. Thereafter it flows towards south-east, turns towards south-west and is joined by another spring fed tributary flowing in from the western part of its catchment. Downstream of this confluence, the Gola river takes a U-turn, flows for a short distance in a westerly direction before turning towards south-east and entering the plains.

The upper course of the Gola river is along a steep gradient. It has cut a gorge across the Siwalik hills. Sub-tropical forests cover its upper and middle catchment while sal and riverine forests are found along its lower course. These forests are under heavy biotic pressure.

LAKES

Kumaun is known as the land of lakes. It has many beautiful lakes, particularly in the hills around Nainital. Important lakes of Kumaun have been described below:

1. Bhimtal Lake

The Bhimtal lake is the second largest lake in Kumaun. It is situated to the north-east of Haldwani in the lower Himalaya. Bhimtal is a very beautiful lake that is visited by thousands of people each year. A small township has come up along the periphery of this water-body. There is an island in the middle of the lake.

Geologists are of the opinion that the lake was created by a series of faults caused by movements in the earth's crust. Surface drainage was thus impeded and this led to the formation of the Bhimtal lake.

Thick forests cover the hill slopes around the Bhimtal lake. These included chir pine, ban oak and mixed deciduous forests. The climatic conditions vary from tropical to sub-tropical. Very heavy rainfall is experienced in this tract during the monsoon season.

The Bhimtal lake is undergoing retrogressive ecological changes due to the following causes:

1. A large quantity of sewage including human excreta is emptied into the lake ecosystem.

2. Heavy biotic pressure on the surrounding forests and construction of roads and buildings has led to an increase in the sediment flowing into the Bhimtal lake.

3. The lake water is polluted by thousands of tourists who visit it each year.

These causative factors are creating the following problems:

—the level of dissolved oxygen in the waters has dropped down to alarming levels.
—there has been an increase in the NO_3-N level in the lake water.
—the fish mortality in Bhimtal lake has increased over the past 30 years.
—the addition of organic matter reduces the drinking water quality of the lake water.

2. Khurpatal Lake

This is one of the many lakes in the Nainital area. It is located near Nainital town and lies in a depression formed by movements in the earth's crust that blocked the flow if a small stream. Steep slopes surround this water body.

Ecological degradation has begun in this lake due to the following factors:

1. Excessive inflow of sediments and sewage from the surrounding areas.
2. Pollution caused by the dumping of solid wastes into the water.

3. Nainital Lake

The Nainital lake is situated in a large depression in the lower hills of Kumaun. The town of Nainital has come up around this lake. The Nainital lake is surrounded by mountain slopes covered with chir pine and deciduous forests. Geologists believe that this lake was formed due to faults produced by movements in the earth's crust. The lake is fed by streams bringing in fresh water from the surrounding mountain slopes.

The locals call the northern part of the lake as Mallital and southern part as Tallital. A road runs around the entire periphery of this lake. The temple of Naina Devi is situated on its banks.

Thousands of tourists visit Nainital each year. They enjoy its serene beauty. Boating, sailing and walking along the lake front is a popular mode of recreation with the tourists.

Ecological degradation has set in the Nainital lake in the recent past due to the causative factors discussed below:

1. A large quantity of sewage including human excreta is emptied into the Nainital lake. The quantity of sewage poured into the lake water is very high in the summer season due to the abrupt increase in the number of tourists staying at Nainital.

2. Accelerated erosion on the slopes surrounding this lake has led to a drastic increase in the inflow of sediments into the lake water. Construction of roads and hotels in the catchment area has added to this problem.

3. Boating and sailing is very popular with tourists. This has disturbed the ecological balance of the lake water.

These factors have led to the following ecological problems in the lake water:

—there has been a drop in the level of dissolved oxygen in the lake waters. This is dangerous for the fresh water fishes living in it.
—there is an increase in the level of NO_3-N in the water of the Nainital lake.
—the fish mortality rate has gone up in the past 3 to 4 decades.
—the drinking water quality of the lake water has gone down.
—excessive siltation has led to a decrease in the area of the lake.

4. Naukuchiatal Lake

This is a small lake in the lower Himalayan hills of Nainital district. It has formed in a small depression due to geological movements in the earth's crust. The slopes around this water body are degraded. As a result, heavy inflow of sediments has endangered the very existence of this water body.

5. Sageriyatal Lake

This is a small lake near the lower Himalayan hills near Nainital. Its depression has been formed by a geological fault. In the recent past ecological degradation has set in due to heavy sedimentation and pollution. This has altered the ecosystem of this water body.

6. Sukhatal Lake

This is one of the many lakes of Nainital area. It is located in a depression near Nainital town. In the past it contained water all round the year but heavy sediment inflow from the surrounding slopes has virtually filled up this lake and hence the name Sukha 'dry' tal. At present there is water only in the rainy season.

GLACIERS

Hundreds of glaciers of various sizes are tenanted in the slopes of the main Himalayan range that stretches across northern Kumaun. A number of these lie on the border of Garhwal and Kumaun.

1. Bagani Glacier

This is a small glacier situated on the slopes of the Nanda Devi massif. High snow-clad peaks surround this glacier. Its melt waters flow into the Rishiganga river which is a part of the Alaknanda drainage system of Garhwal.

2. Burh Glacier

The Burh glacier is located on the lower slopes of the Nanda Devi massif along the border of Garhwal and Kumaun. It has a length of about 3 kms. The Panwali river rises from the snow melt waters of this glacier. It is a tributary of the Sunderdhunga river which in turn drains into the Alaknanda river system of Garhwal.

Small tributary glaciers join the Burh glacier. These are:

—glaciers in hanging valleys descending down from the slopes of the Panwali doar and Bauljuri peaks.

—small glaciers descending down from the Mangtoli peak.

A vast quantity of fluvo-glacial sediments occurs on the floor of the glacial trough. The channel formed by the snow melt waters displays a braided pattern. The glacial trough narrows down to an epigenetic gorge about 3 kms downstream of the snout of the Burh glacier.

Avalanches are common particularly in winter. Alpine grasses come up on the moraines in late summer and become dormant in late autumn.

3. Burla Glacier

This is a small glacier located in a hanging valley on the western slopes surrounding the Pindari glacier. It is a tributary of the main glacier.

4. Kafni Glacier

The Kafni glacier is situated on the south-western slopes of Nanda Devi massif along the border of Garhwal and Kumaun. Its melt waters give rise to the Kafni river which is a tributary of the Pindar river that in turn flows into the Alaknanda river system.

Two main tributary glaciers feed the Kafni glacier. These are: (a) a valley glacier on the eastern flank of its glacial trough, and (b) a small glacier on the western slope of its trough.

The Kafni glacier has a length of about 2.5 kms though its trough is about 5 kms long. It has a width of about 1 km in the upper portion. The glacier has receded by about 2 km in the past few decades.

High mountain ranges encircle the Kafni trough on two sides. Its side walls are very steep and bear small glaciers in hanging valleys. The crest of the terminal moraine is at a relatively lower level. A vast quantity of debris and fluvo-glacial sediments are found on the valley floor.

The melt water forms a braided, channel on the valley bottom. Alpine vegetation comes up in the lower trough from late summer to mid-autumn.

5. Mrigthuni Glacier

This glacier is located on the slopes of the Nanda Devi massif on the water divide of Garhwal and Kumaun. It has a length of about 5 kms and is tenanted in a cirque on the slopes of the Mrigthuni peak. The Sukhram river rises from the snout of this glacier. It is a tributary of the Sunderdhunga river.

The glacial trough is known as Sukhram trough. It is strewn with thick deposits of fluvo-glacial sediments.

6. Pindari Glacier

This is the largest glacier of Kumaun. It is situated in a huge amphitheatre formed by high peaks amongst which the Nanda Kot is the most prominent. The Pindari glacier gives rise to the Pindar river which is a part of Alaknand a river system.

The Pindari glacier gathers snow from three feeder glaciers. These are:

— a large glacier from the slopes of the Nanda Kot peak in the north-west.
— the Burla glacier in a hanging valley on the western slopes.
— a smaller glacier in the east.

The glacier is located in a wide trough. Its eastern flank is very steep. The trough narrows down near the snout of the glacier. This glacier has retreated by about 3.2 kms in the past 150 years. It has left behind a vast wasteland consisting of boulders and recently laid down fluvo-glacial sediments.

5

Geology

Kumaun has an intricate lithology and structure which marks the transition between the geology of the western and central Himalaya. The main difference between the geology of these two areas is that the trans-Himalayan or tethyan zone is extensively developed in Jammu and Kashmir and H.P. and is less extensive in Kumaun and the central Himalaya.

On the basis of lithology, structure and tectonics, Kumaun may be grouped into the following four geological zones (from south to north). These more or less coincide with the physiographic zones of Kumaun discussed in the first chapter.

1. Siwalik or Outer Himalaya
2. Lesser or Lower Himalaya
3. Central or Higher or Main Himalaya
4. Trans or Tethyan Himalaya.

The lithology and structure of each of these zones has been discussed in this chapter.

Siwalik or Outer Himalaya

The Siwalik hills run more or less in a NNW-SSE direction in the southern tract of Kumaun. The Siwalik or outer Himalaya of Kumaun is made up of a number of sub-units, viz.

—the terai and bhabar tract which is made up of a thick deposit of recent to sub-recent sediments

—the Siwalik hills that are comprised of the rocks of the Siwalik succession. These have been discussed in detail in the following text.

—the dun valleys that are made up of a thick deposit of recent to sub-recent sediments, mainly the dun gravels.

The succession of the Siwalik sequence of Kumaun is given below:

Upper Siwalik	3. Boulder conglomerate
	2. Pinjor stage
	1. Tatrot stage
Middle Siwalik	2. Dhok Pathan stage
	1. Nagri stage
Lower Siwalik	2. Chinji stage
	1. Kamlial stage

Kamlial Stage

This is the lower most stage of the Siwalik succession in Kumaun and other parts of the western Himalaya. The main characters of this stage are outlined below:

—it consists of hard and compact, red and grey sandstones and purple shales.
—these may also be accompanied by bands of pseudo-conglomerates
—the sandstones occur in a thick sequence and individual beds may be more than a metre in thickness.

Chinji Stage

This is the topmost stage or horizon of the lower Siwaliks. The main characters of this stage are outlined below:

—it is made up of bright red to maroon shales and sandstones
—these are accompanied by intercalations of silt stones that may be of varying colours
—the thickness of the Chinji stage ranges from about 300 to 800 mts
—this stage bears a rich assemblage of fossils.

Nagri Stage

This is the lowermost stage of the middle Siwalik rock sequence. The main characters of the Nagri stage are outlined below:

—it is comprised of hard and compact grey sand stones whose colour may vary from buff to brown in certain sections

—these sandstones are associated with grey, green or muddy shales
—the Nagri stage is relatively poor in fossils particularly those of mammals.

Dhok Pathan Stage

This horizon overlies the Nagri stage and is the topmost stage of the middle Siwalik sequence. The main characters of the Dhok Pathan stage are outlined below:

—it is composed of brownish sandstones, gravel beds, drab shales, orange clays and claystones
—this horizon is about 1000 mts. in thickness
—the Dhok Pathan stage is the richest stage in fossils in the entire Siwalik rock sequence. It contains both animal and plant fossils.

Tatrot Stage

This is the lowermost horizon of the upper Siwalik rock sequence. It overlies the Dhok Pathan stage of the middle Siwaliks. The main characters of the Tatrot stage are outlined below:

—it is made up of soft grey sandstones that attain a brownish shade in certain sections.
—these are associated with drab and brown clays and conglomerates.
—there occurs a coarse conglomeratic horizon at the base of this stage. The presence of this horizon shows evidences of quick deposition by rivers and delta like structures laid down under heavy rainfall conditions.
—the Tatrot stage too is rich in fossils.

Pinjor Stage

This stage overlies the Tatrot stage and occupies the middle part of the upper Siwalik sequence. The main characters of the Pinjor stage are outlined below:

—it is made up of coarse grits, conglomerates and sandstones
—there may also occur variegated sands and pink coloured silts. The pink colour is indicative of relatively drier conditions of depostion of this stage
—the Pinjor stage has a thickness of upto 150 mts

—it bears a rich assemblage of fossils of mammals, reptiles, birds and plants.

Boulder Conglomerates

This is the topmost horizon of the Siwalik rock sequence. The main characters of the Boulder conglomerate horizon are outlined in the following text:

—it consists of a thick horizon of conglomerates. They are mainly from the lesser and main Himalayan formations which lies towards north. These sediments are made up of igneous, sedimentary and low grade metamorphic rock conglomerates
—in certain sections the aeolain material deposited under dry conditions is found in between the boulder conglomerates.

The different authors have given sequences of the Siwalik rocks in Kumaun and adjoining areas. A comparative statement is given in the following table:

	Authors		
	Pilgrim (1934)	Colbertt (1935)	Lewis (1937)
Upper Pleistocene	–	–	Tawi
Middle Pleistocene	–	–	Tawi
Lower Pleistocene	Boulder conglomerate	Boulder conglomerate Pinjor	Break
Upper Pliocene	Pinjor	Tatrot	Pinjor-Tatrot
Middle Pliocene	Tatrot	Dhok Pathan Nagri	Dhok Pathan
Lower Pliocene	Dhok Pathan	Chinji	Nagri
Upper Miocene	Nagri	Kamlial	Chinji
Middle Miocene	Chinji Kamlial	–	Kamlial
Lower Miocene	Base of the Siwalik Sequence		

Structure and Tectonics

Rocks of the Siwalik sequence display an intricate structural set up. The salient features of the structure of the Siwaliks are:

1. The rocks are largely folded into a series of broad, open anticlines and synclines.

2. These structures have been evolved during the later phases of

Himalayan orogeny.

3. The crest of the anticline overlaps the crest of the Siwalik ridge. This has been fractured by a number of faults and their offshoots.

4. The Siwalik rock sequence forms a gently dipping syncline in the dun valley which has been filled up by recent sediments or dun gravels.

5. These rocks are separated from the lesser Himalayan rock sequence by the main boundary thrust which has terminated their northern extension.

6. Overthrusts of local extent occur in the Siwalik rock sequence.

7. In the south a fault separates the Siwalik rocks from the sediments of the Ganga plain.

Conditions of Deposition

The coarse and often poorly graded sediments of the Siwaliks suggest that they have been carried by fast flowing and large bodies of water and deposited in wide, shallow water depressions in swampy basins. Alternations of coarse and fine sediments suggests that seasonal deposition took place.

The coarser materials were deposited during floods (rainy season) and finer sediments under drier conditions.

There is an extraordinary similarity in the nature of such rocks along the strike which suggests the same depositional basin and similarity of source rocks. Moreover, experts are of the opinion that the Siwalik sediments extend deep down below the Ganga plain over an area of hundreds of square kilometres.

Climatic Conditions

A warm and humid climate prevailed during the greater part of the sedimentation of the Siwaliks. The lower Siwalik sequence and the Nagri stage of the middle Siwaliks are believed to have been deposited under moister climatic conditions. Humidity probably decreased when sediments forming the Dhok Pathan stage were laid down and sedimentation took place in partly marshy conditions and partly on dry land.

During the deposition of the middle Siwalik sequence, the depositional basin was shifted further towards south as a result of tectonic activity. However, wetter conditions returned once again during the period when the upper Siwaliks were laid down. The climate clearly became cooler during the deposition of the post Tatrot sediments when another upliftment took place. During this period the animals that roamed in this area either died or migrated away due to the extreme cold conditions prevailing in this tract.

Lesser or Lower Himalaya

In 1934, Auden gave the first detailed account of the geology of the lesser Himalaya of Garhwal and Kumaun. The stratigraphic succession of the rocks of Kumaun based on the works of different authors is given below:

Group	*Formation*	*Member/Bed*
KUMAUN	Subathu	Dagshai
		Nummulitic
	—UNCONFORMITY—	
	Tal	Upper Tal
		Lower Tal
	Krol	Krol E
		Krol D
		Krol C
		Krol B
		Krol A
	Blaini	
	Binj	
	—UNCONFORMITY—	
	—CHAIL NAPPE UNIT—	
	Dudatoli-Almora crystallines	
	Manila	
	Nagthat	
	Chandpur	
	Saknidhar	

The main boundary thrust separates the rocks of the lesser Himalaya of Kumaun from from those of the Siwalik sequence in the south. In the north these are cut off from the rocks of the main Himalaya by the main central thrust.

Saknidhar Formation

This is the oldest formation of the Kumaun group. It is made up of limestones, shales and quartzites. There may occur intercalations of basic volcanic rocks at some places. The Amritpur granite has intruded into the rocks of this formation in Nainital.

Chandpur Formation

This formation overlies the Saknidhar formation. It consists of a thick sequence of dark grey, green, maroon and pure green phyllites, slates and shales.

Nagthat Formation

The Nagthat formation overlies the Chandpurs. It is made up of a thick sequence of orthoquartzite with subordinate purple coloured phyllites. This rock formation has also been referred to as the Maithana quartizite in Garhwal. It may be divided into two:

a) Devitonk-Khirsu quartzite
b) Pabo phyllite.

Within the quartzitic band there is a gradual decrease in the size of individual grains from coarse, gritty and occasionally pebbly in eastern Kumaun to fine grained near the western border with Garhwal.

Manila Formation

The Nagthat quartzites are conformably overlain by a thick sequence of rocks consisting of phyllites of varying colours associated with subordinate quartzite. It has been referred to as the Bamsyun formation in Ranikhet area. In eastern Kumaun, this formation appears to merge with the Lohaghat phyllites and are intruded by the Champawat granodiorite.

Dudatoli-Almora Crystallines

Rocks of this formation overlie the Manila phyllites. They are made up of:

a) *Dudatoli schist:* Primarily consisting of metamorphosed pelites, semipelites and psammitic sediments which have been rhythmatically laid down. These have been deposited in a normal squence.

b) *Dudatoli granite*: Primarily consisting of granite and granite gneiss. Largely a heterogenous association of metamorphics, migmatites and granite-gneiss.

Chail Nappe Unites

The rocks of the Chail nappe units overlie the other rock units of

Kumaun in many sections. Fuchs and Sinha (1978) have recognised three nappe zones in this region, viz. Chail C_1, Chail C_2 and Chail C_3. According to them, "The Chail Nappe, which at the base of the Chor crystallines may be traced to the north is the highest unit (C_3) of the Chail nappes. The Deobans and associated Chails (phyllites and quartzites), underlying this highest nappe form lower subsidiary units of the Chail nappe system."

In Kumaun the famous Berinag quartzites are considered to be a member of Chail 3. South of Baijnath this unit comprises of intrusive granite-gneiss, metavolcanics etc. In the Pithoragarh area the Ladhiya formation and Betalghat formation are also a part of this nappe.

Binj Formation

This formation has also been referred to as the Shankarpur formation. It is the lowest formation of the Carboniferous to Tertiary succession. It unconformably overlies the Binj and Chandpur formations of Garhwal and Kumaun. This is unconformably overlain by the Blaini boulder beds.

The Binj formation has further been sub-divided into two:

(a) upper Binj unit or Bansi member
(b) lower Binj unit or Jogira member

The Binj formation is made up of slates, phyllites and shales of varying colours.

Blaini Formation

The Blaini formation overlies the Binj formation. It is made up of the following sequence:

—glauconitic sandstone
—boulder or conglomerate beds consisting of boulders and/conglomerates of purple and green coloured shales, dolomite, limestone, quartzite, slates, phyllites and even igneous rocks embedded in a matrix.
—boulder slate (found in some areas).

The presence of this formation is indicative of a break or a series of breaks in the deposition of sediments and conditions of flooding. These boulder beds can be correlated with boulder beds of the Permo-Carboniferous age found elsewhere in the Himalaya.

Krol Formation

This is a well developed rock formation found in the lesser Himalayan tract of Kumaun. It shows best development in the Nainital area. The Krol formation is made up of thick deposits of the following members or units:

Member	*Lithology*
Krol E:	made up primarily of agrillaceous limestone and shale (in a few sections)
Krol D:	consists of sulphurous cherty limestone and occasional pockets of barite.
Krol C:	this member consists of an extremely thick sequence of limestone which is occasionally mixed with dolomite and marble.
Krol B:	made up primarily of red, maroon and green shales and sporadic occurrences of deposits of gypsum.
Krol A:	consists mainly of calcilutite, limestone and calcerous slate. Lenses of gypsum may be found occasionally.
Infra Krol:	comprises mainly of grey shales, slates, graywackes and siltstones.

Tal Formation

This is another well developed rock sequence which overlies the Krol E member of the Krol formation. The Tals are divided into two members or units:

a) *Upper Tal*: It consists mainly of a sequence of quartzite with limestone as an associate. This unit is seen to the north of Nainital. Very clear pebbly horizons are seen in the upper Tals in a number of sections. This is indicative of a break in deposition of the siliceous sediments.

In Kumaun, the upper Tals have also been referred to as the Phulchatti quartzite and the Mainkot shell limestone.

b) *Lower Tal*: The lower Tal unit too has attained best development in the tract to the north of Nainital. It is made up of purple, red, maroon, black and green coloured shales and phyllites which are thinly bedded.

Conodonts and other remains of micro organism are also found in this horizon. These include *Oneotodus erectus* and *Palfodus* sp.

Subathu Group

Rocks of this group unconformably overlie the upper Tal rocks in narrow bands in some sections of Kumaun. They consist of shales, slates and phyllitic sequences. This group is made up of two formations, viz.

(a) the Subathu formation
(b) the Dagshai formation

Structure and Tectonics

On the basis of structure and tectonics the lesser Himalayan zone of Kumaun may be sub-divided into two NN-SE trending structural (tectonic zones). Each of these zones is bounded by lines of tectonic disturbance.

Garhwal Tectonic Zone

This is the northernmost tectonic zone of the lesser Himalaya of Kumaun. It is well developed in Garhwal and relatively less prominent in Kumaun. In the north is the main central thrust which forms the boundary of the lesser and central Himalaya while the north Almora thrust forms the southern limit of the Garhwal tectonic zone.

The following three phases of dislocation have been recognised in this tectonic zone.

First phase: Development of tight, isoclinal plunging folds with plunge towards NW or SE. Chevron like folds have also formed. Folds of this phase may be overturned or doubly plunging folds.

Second phase: Development of open wraps or folds with plunging anticlines and synclines towards north or south. N-S faults have also developed.

Third phase: Development of broad, open plunging folds which are superimposed on the folds of the earlier phases. Such folds may also be developed on the limbs of the 1st or 2nd phase folds. There has also occurred the development of large scale folds in this phase.

A number of outliers are found in the Garhwal tectonic zone. These occur at Askot, Chipalkot, Lekh, Baijnath and Dharamgarh.

Kumaun Tectonic Zone

This tectonic zone lies between the Garhwal tectonic zone in the north and the Siwalik or outer Himalaya in the south. Its northern boundary is marked by the north Almora thrust while in the south the boundary is formed by the main boundary thrust. Two rock sequences or successions lie within this tectonic zone. These are:

—the Kumaun group (Pre-cambrian and Cambrian)
—the Upper Palaeozoic to lower Tertiary sequence (Binj, Blaini, Krol, Tal, Subathu-Dagshai formations).

These rock successions contain two nappe systems: the Krol nappe and the Garhwal nappe which have been thrust over the basement of the Shimla slates and Nummulites. However there may occur a strong angular unconformity between the Saknidhar formation and Blainis which poses doubts over the status of the Krol and Dudatoli-Almora thrust.

In the Nainital area, the Blainis rest unconformably over the rock sequence which is an eastern extension of the Saknidhar formation. The Krol thrust is not present in this area and the rocks of the Krol belt are thus autochthonous.

After the Almora crystallines there occurs a continuous normal rock sequence in the Kumaun group, viz. the Saknidhar formation, Chandpur phyllite, Nagthat quartzite, Manila phyllite etc. The metamorphic boundaries are at an angle to the lithologic ones.

In Paithani area the Manila phyllites have been metamorphosed to garnet-mica-schist. This has probably led to the belief that there is a thrust plane below the Almora-nappe. Another prominent dislocation of this zone is the south Almora thrust. This is a large strike slip fault that off-sets many formations of Kumaun. In eastern Kumaun this thrust brings the rocks of the Manila phyllites to rest over the Saknidhar formation thus establishing the autochthonous nature of the Dudatoli-Almora crystallines.

The following three distinct tectonic units occur in Kumaun area.

1. *Parautochthonous unit*: Consisting of the Kumaun group and the overlying upper Palaeozoic to upper Tertiaries.

2. *Chail nappes*: Made up of three clear-cut nappe units C_1, C_2 and C_3 from south to north. They account for a large part of the central zone formed by the Garhwal and Kumaun tectonic zones.

3. *Crystallines nappes*: Consists mainly of the crystalline rocks that occur in the northern part of the lesser Himalayan zone viz. Dudatoli-Almora crystallines.

Central or Main or Higher Himalaya

This geological zone lies to the north of the lesser Himalayan zone. It is separated from the latter by a deep seated tectonic linement known as the main central thrust. In the north this zone includes the rocks of the main Himalayan mountain wall and extends till the tethyan or trans-Himalayan zone.

The principal rock formation/succession of the central or main Himalaya of Kumaun have been discussed in the following text:

Amphibolitic Sills

Sills of amphibolitic rocks having gabbrodioritic composition occur to the north of the main central thrust. Well developed outcrops of the main central thrust zone are found in the Kali valley in eastern Kumaun. This thrust occurs just north of the township of Dharchula where it has an angle varying from 10 to 15 degrees.

Gneiss and Quartzite

These are metamorphic rocks that either overlie the amphibolitic sills or are in direct contact with the main central thrust. It is made up of:

a) Biotite-alkali felspar gneiss with rare occurrences of plagioclase grains.

b) Muscovite-biotite augen gneiss with large amethyst coloured augens of quartz. The felspars are broken and sheared in such a rock type while the amethyst is more or less rounded indicating a quartz prophyry.

c) Biotite schist occurs in a sheared zone.

d) Biotite-granite gneiss is found in most sections. It has a predominance of alkali felspars.

e) Biotite-sericite schists with intercalations of biotite-alkali felspar prophyroblastic gneiss having bluish grey orthoclases. There have developed beautiful myrmekitic reaction zones of plagioclase-andesine around the arthoclase grains.

f) White quartzite bands.

Sedimentary Section of Sirdang

The rocks of this zone are also known as the Sirdang zone. They overlie the gneissic and quartzitic rocks discussed in the above text. The sedimentary section of Sirdang consists of the following rock types:

a) Dark grey to blackish phyllite slates having thin intercalations of marbles. They are graphitic in content and are free of lime. These slates also contain small scales of graphite-sericite. In the upper portions the slates are intercalated with muscovite bearing calcium marbles.

b) Amphibolitic horizon.

c) Pure white sericite quartzites follow the amphibolite horizon. The quartzites are very well developed in the upper valley of the Kali river.

Further towards west this zone can be connected to the northern Tejam zone and the lower thrust mass actually pinches out in the Goriganga valley where it is well exposed around Munisiari.

A generalised succession of the rock sequence of the main Himalaya in the Kali valley is as under:

i) biotic phyllites (graphitic and calcerous)—Sirdang zone
ii) pure white sericite marble—Sirdang zone
iii) white quartzites
iv) dioritic amphibolite
v) biotite-muscovite gneiss
vi) quartzites and lime silicates (Garbyang and Budhi schists)

Quartzites and Lime Silicates

In the biotite-muscovite gneiss and psammite gneiss, there occur gradually the first few thin horizons of lime horizons which are rich in garnet, zoisite and calcite and calcite content. The lime silicates are followed by a fairly thick horizon of sericite quartzites which contain exceptionally large grains of kyanite with garnet and tourmaline.

The upper part of these quartzites have been tightly folded and within these begin to occur the first of the large pegmatite dykes. Such dykes comprise of tourmaline pegmatite which is poor in mica content and is associated with dykes and small stocks of fine grained granite rich in muscovite. These granitic bodies have been termed as muscovite-tourmaline-aplite granite. Towards Api peak, there occur well stratified and finally subfolded intercalations of biotite psammitic gneiss.

Budhi Schists

The rock sequence described in the above text grades into a band of well developed schists—Budhi schists of Budhi prophyroblastic schists. There is a decrease in the occurrence of dykes in the Budhi schists. Within these schists there occur large transverse biotites which are having sieve like intrusions of quartz. It appears that the latter was originally parallel to the direction of schistosity but were rotated with the biotite grains by late orgenic movements.

The Budhi schists represent a transition from the highly metamorphic rocks to the slightly metamorphosed rocks of the pre-Cambrian to early Cambrian sediments, i.e. Martoli, Ralam and Garbyang formations.

Lower Sedimentary Belt

This belt comprises of rocks of the normal sedimentary cover of the crystalline thrust sheet of the north-central Himalaya of Kumaun.

Martoli Formation

This formation conformably overlies the Budhi schists. They are well exposed in the Goriganga valley, near the Nanda Devi massif. In this section, the Martoli formation has a thickness of over 3000 mts. This formation comprises mainly of calcerous phyllites in which there occur fine grained layers of quartzite.

Within the Martoli formation, there occur a coarse type of conglomerate composed of elongated pebbles of pinkish quartzite which may be over 30 cms in diameter and are found to be embedded in a matrix of greenish quartzite. These conglomerates have been named as the Nandakot conglomerates.

Ralam Formation

This formation unconformably overlies the rocks of the Martoli formation. An angular unconformity is assumed to have occurred around the base of the Cambrians. Thus, the overlying Garbyang formation is said to have an Cambrian age. The main rocks of the Ralam formation are comprised of rounded pebbles of black to red quartzite, which are embedded in a matrix of graywacke—the Ralam conglomerate. These are different from the Nandakot conglomerates discussed above.

Garbyang Formation

The Garbyang formation comprises of:

—greyish green to purple, fine quartzites and weathered orange dolomite horizons;
—very fine grained, yellow calcerous sandstones;
—calcerous, argillaceous banded silts;
—calcerous, quartzitic slates and phyllites.

As has been discussed in the above text, the Garbyang formation is of Cambrian age. However, there is no clear cut demarcation between this formation and the underlying Martoli formation, under which occurs the lower base of Cambrian unconformity. Here, it may be pointed out that such sedimentary rocks represent the pre-Cambrian-Cambrian transition in other parts of the Himalaya also.

Tethyan Himalaya

This zone lies in the area to the north of the higher Himalaya. In Garhwal-Kumaun, the rocks of this zone are not very extensively developed as compared to the J & K and H.P. Himalaya. Moreover, the rocks of the tethyan zone occur more widely in Kumaun, than in Garhwal, where they are found in sporadic patches. The following table shows the success of tethyan rocks in Kumaun:

Age	*System/Series/Formation*
Cretaceous	Cretaceous flysch
Jurassic	Laptal series
Triassic	Kioto limestone
	Kuti shale
	Kalapani limestone
	Chocolate series
Permian	Kuling shales
Middle to upper Devonian	Muth series
Carboniferous	Kali series
Silurian	Variegated series
Ordovician	Shiala series
Cambrian	Garbyang series
Pre-Cambrian	Vaikrita system

Vaikrita System

This system is of pre-Cambrian age. It comprises mainly of highly folded mica-schists, slates, phyllites etc. They have been so named by Griesbach (1891). They rest upon gneissic formations and their position is controversial as to whether they are to be placed in the higher Himalaya or the tethyan Himalaya.

Garbyang Series

In Kumaun, the equivalents of the Haimantas are the Garbyang series, named by Heim and Gansser after a village of the name in Kali valley. They are well exposed from Nanda Devi peaks region in the west to Nampa peaks in west. The Garbyang series consist mainly of phyllites, slates, fine grained calcerous sandstones, argillaceous dolomite (which contain green chloritic bands, these have weathered to a brownish colour). It appears that these chloritic bands are partially metamorphosed basic tuffs.

The Garbyang series is best developed in the valley of the river Kali, where it attains a thickness of over 1000 mts. Poorly preserved, flattened, gastropod shells occur in the phyllites which points to a Cambrian age of this series.

Shiala Series

The Shiala series lies sandwiched between the Garbyang series and Silurian shales. It has been named after Shiala pass, where it attains a thickness of over 400 mts. Main rock types include shales intermixed with grey and greenish marly limestone, containing at times crinoidal fragments.

The following main fossils indicating an Orovician age for the Shiala series have been reported:

Calymene douvillei
Orthotetes pecten
Rafinesquina subdeltoidea
Orthis thakil
O. orbignyi
Sowerbyella sp.

Variegated Series

This series comprises of rocks of Silurian age. It comprises of alternations of limestones and marls and siliceous shales of varying colours. The red colour is very dominant in these shales and hence they have also been named as the red Silurian shales.

The rocks of this series are well developed near Shiala pass; at Gunji where the siliceous shales overlie the Garbyang series with disconformity.

Muth Series

The Muth series comprises mainly of hard, white massive quartzites, accompanied by dolomitic layers, which have weathered to produce a brown colour. The quartzites have been named as the 'Muth quartzites'.

In certain sections, the Muth series is accompanied by a dark limestone, which bears the following important fossils:

Atrypa reticularis	*A. aspera*
Camarophoria phillipsi	*Favosites* sp.
Orthis bistriata	*Orthoceras* sp.
Pentamerus sublinguifer	*Rhynchonella omega*

On the basis of these fossils, a middle to upper Devonian age has been assigned to the Muth series.

Kali Series

This series is not very well developed. It is equivalent to the Po, Lipak series and Fenestella shales of Spiti.

Kuling Shales

The Muth series are conformably overlain by the black coloured Kuling shales of Permian age with rocks of the Kali series occurring sporadically in between. The contact between these two is very sharp and represents a break from the top of Devonian times to lower Permian. Basal conglomeratic horizons which underlie the Kuling shales in Spiti area are absent in Kumaun. The carboniferous gap is filled by sediments resembling the Po and Lipak series of Spiti (Kali series), in a number of places (these are absent in a majority of the sections). The Kuling shales, which have also been termed as the Productus shales, represent a deep marine deposit. They are, by and large, poorly fossiliferous. The important ones being:

Cyclolobus walkeri	*C. oldhami*
Productus himalayanus	*P. semireticulatus*
P. abichi	*Spirifer nitiensis*
S. tibetanus	

Exotic Facies

These are huge masses of sedimentary rocks which have virtually no similarity to the surrounding country rock. They vary in size from boulders to hillocks and even sheet like masses in certain cases. They are generally of Permian and Mesozoic age and are found around Kiogar-Chitichum in the north-western tip of Kumaun.

A controversy exists regarding the origin of these 'exotics' as they have no relation whatsoever with the surrounding country rocks. One school of thought is of the opinion that these are parts of Tibetan formations and have been transported by lava flows to their present position. Another hypothesis advanced by Diener considers them to be remnants of huge thrust sheets or nappes from north.

Lithologically, these comprise of red, pink, maroon and white limestones of all ages between Permian to Cretaceous. They are dominant with the following fossils:

Cheiropyge himalayensis
Cyclolobus walkeri
Clisiophyllum
Dielasma elongatum
Maritima elegans
M. glabara
Productus abichi
P. chitichunesis
P. semireticulatus
P. gratiosus
Spirifer wynnei
S. fasciger
S. tibetanus
Spirigeralla grandis
S. xenapsis sp.
Zaphrentis beyrichi

Triassic Succession

The Triassic succession comprises of the following:

Chocolate Series

A 30 to 50 mts thick band of clay-iron stone layers with intercalations of shales. These are passing into bands of dark shales. This series contains the following important fossils:

Meekoceras hodgsoni
Ophiceras demissum
Pseudosageceras sp.
Vishnuites pralambha

The age of this series has been placed as Scythian.

Kalapani Limestone

Overlying the chocolate series is a 20 to 50 mts. thick band of limestone—the Kalapani limestone. It comprises of three horizons:

a) *Upper horoizon*—well bedded limestone with brachiopods and corals.
b) *Middle horizon*—a hematitic layer rich in ammonites.
c) *Lower horizon* —a limestone horizon rich in tropites.

The following are some important fossil occurrences:

Arcestes sp.
Halobia sp.
Sirentes sp.
Gymnites sp.
Halorites sp.
Thisbites sp.

On basis of these fossils an age varying from Anisic to Carnic has been assigned to the Kalapani limestone.

Kuti Shale

These are largely micaceous shales with calcerous flanks, having a thickness of over 100 mts. Fossils are generally rare, the important ones being:

Antomites sp.
Halorites procyon
Placites sakuntala
Discophyllites sp.
Juvavites sp.
Thetidites buxleyl

This entire group is fossils of Noric age.

Kioto Limestone

This is a fairly well developed limestone belt which`extends from the north-eastern portion of Kumaun, right upto Garhwal. It comprises of dark blue, well bedded limestones with oolites in the lower horizon. The age of this limestone is of upper Triassic and Rhaetic. The only fossils of significance, which occur in the Kioto limestone are some bivalves.

Laptal Series

The Kioto limestone is succeeded by the Laptal series in northern and north-western Kumaun. This series comprises of brown sandy limestones, dark marly limestone, yellow and red spotted calcerous horizons containing fossils of Belemnites and bivalves. The Laptal series are of Jurassic age.

Cretaceous Flysch

In northern Kumaun-Garhwal area, the Cretaceous are represented by the flysch sediments. They are:

a) Upper flysch—upper Cretaceous age
b) Lower flysch (Guimal sandstone)—lower Cretaceous age.

By and large, fossils are rare except some ammonites, bivalves etc. Heim and Gansser have recognized the following sub-divisions in the upper flysch:

3. black slaty shales with Fucoids;
2. purple marly shales bearing forminifera.
1. green shales with sandstone layers.

6

Natural Vegetation and Forests

Nature has bestowed Kumaun with rich forests that are made up of thousands of species of trees, shrubs, herbs and climbers. These combine to form forest types which vary from the sub-tropical forests in the lower hills to the dense temperate forests and the dry vegetation of the upper Darma valley of northern Kumaun.

The nature and type of vegetation occurring in the particular part of this region depends upon a combination of the factors listed below:

a) prevailing climatic conditions
b) altitude
c) aspect, topography and slope
d) edaphic and biotic factors
e) latitude

FOREST REGIONS

The following forest regions may be delineated in Kumaun (after Puri, 1960 and Negi, 1991).

1. Sub-montane or Sub-tropical Region

This forest region is made up of the following areas:

a) Terai and bhabar tract
b) Dun valleys
c) Lower and middle Himalaya upto 1500 mts
d) Siwalik hills.

Broadly speaking this forest region includes the tracts lying below an elevation of about 1500 mts. Summers are very hot in this region and the mercury may soar to over 40°C. The SW monsoons account for bulk of the total annual precipitation. The monsoon rains begin in early July and continue till the middle of September. They bring relief from the scroching summer heat and from the hot and dusty winds that affect the terai and bhabar tracts. Autumn is fair and sunny, winters are mild and frost commonly occurs in this season.

The main forests occurring in this region are listed in the following text:

a) Sal forests dominated at different elevations by *Shorea robusta.*
b) Chir pine forests dominated by *Pinus roxburghii.*
c) Khair and sisham forests dominated by *Acacia catechu* and *Dalbergia sissoo*
d) Lower deciduous forests consisting of a number of dry deciduous species.

2. Montane or Temperate Region

This forest region extends from an elevation of about 1500 mts to 3500 mts in the middle and higher Himalayan hills of Kumaun. It may further be sub-divided into: (a) lower temperate forest region, and (b) upper temperate forest region.

This tract experiences a typical temperate climate. Very heavy rainfall is received in the lower regions while the upper reaches experience heavy snowfall. Winters are long and severe and the mercury may drop down to levels below the freezing point.

The main forests occurring in this forest region are:

a) Oak forests dominated at differenct elevations by *Quercus leucotricophora; Q. himalayana* and *Q. semecarpifolia.*
b) Deodar forests dominated by *Cedrus deodara.*
c) Kail forests dominated by *Pinus wallichiana.*
d) Fir and spruce forests dominated by *Picea smithiana* and *Abies pindrow.*
e) Mixed broad leaved forests consisting of *Aesculus indica, Juglans regia, Rhododendron* etc.

3. Sub-Alpine and Alpine Region

This region represents the uppermost limit of tree and vegetative growth. It extends from an elevation of about 3000 mts to the snowline in the following tracts: (a) both slopes of the middle Himalaya, (b) southern slopes of the main Himalaya, and (c) inner dry valleys.

Typical arctic and sub-arctic climatic conditions prevail in this forest region. Summers are short and mild. The snow melts in late spring after which the summer season sets in. Winters are long and severe and very heavy snowfall occurs during this season. The ground is covered by a thick blanket of snow for several weeks at a stretch.

The growing period available to the plants is very short. They tend to develop a vast network of roots which helps to sustain the plants during the cold season. The biomass below the ground is more than that above it. The plants develop special adaptations to live in the extreme cold climate.

The main forests occurring in this region are:

a) Alder forests dominated by *Alnus nitida*
b) Upper fir forest dominated by *Abies spectabilis*
c) Birch forest dominated by *Betula*
d) Rhododendron forest dominated by *Rhodondendron*
e) Moist and dry alpine scrub consisting of alpine flora mainly grasses and shrubs.

4. Trans-Himalayan Region

This region consists of the tracts lying in the rain shadow of the main Himalayan range, viz. parts of northern Pithoragarh (upper Darma valley).

Rainfall is very low in this region and hence thorny species dominate. Tree growth is stunted on exposed slopes. Moist strips of land along channels formed by the snow melt waters bear tree growth.

The main forests occurring in this forest region are:

a) Dry deodar forest dominated by *Cedrus deodara*
b) Dry juniper scrub dominated by *Juniperus*
c) Dry alpine scrub consisting of shrubs and stunted trees just below the snowline.

FOREST TYPES

The forest types found in the Kumaun hills resemble those of the adjoining tracts of Garhwal and western Nepal. The following are the main forest types found in Kumaun. These have been delineated on basis of the parameters listed below (after Champion and Seth, 1968 and Negi, 1990):

a) Nature and composition of the vegetation
b) Prevailing climatic conditions
c) Altitude and latitude
d) Proximity to the snowline
e) Aspect.

Moist Sal Forest

Extensive moist and almost pure forests of sal (*Shorea robusta*) are found in the following physiographic tracts of southern Kumaun:

—on moist northern slopes of the Siwalik hills
—the terai and bhabar tract
—dun valleys
—lower Himalayan foothills.

These forests are dominated by sal which comprises of over 90 per cent of the crop. In Kumaun the moist sal forest is relatively more thick than the similar forests found in Garhwal or south-eastern H.P.

The factors outlined below are responsible for sal being more aggressive and dominant that any of its associates:

a) It has a high coppicing power.
b) It has an ability to regenerate quickly and steadily even under adverse conditions such as heavy grazing, lopping and frequent fires in summer.
c) It is able to readily adapt itself to the local edaphic and climatic conditions.

Sal trees may attain a height of upto 40 mts or more. They have long, cylindrical boles. The sal tree remains leafless for a very short period at the beginning of the summer season. Thus, it is semi-evergreen to evergreen in nature.

In summer, sal forests have a new dense foliage which provides a cooler shade than other deciduous forests of the lower hills that may be devoid of leaves in this season.

The middle and lower storeys of sal forests are well developed particularly in the dun valley and the terai and bhabar tract. Most of the shrubs occurring in the moist sal forests are semi-evergreen. Repeated fires bring about their replacement. Many species of climbers are also associated with these forests. Bamboos and canes may be found in moist shady depressions while chir pine is mixed with sal trees in the upper reaches.

Riverine forests dominated by khair and sisham are found along rivers and streams within the moist sal forest.

The following sub-types of the moist sal forest are found in the southern tract of Kumaun.

a) *Siwalik sal forest*: As the name suggests, this forest is found on the north facing, moist slopes of the Siwalik hills, usually upto an elevation of about 1000 mts. Chir pine is associated with this forest on hill tops.

Moist, sub-tropical to tropical climatic conditions are experienced in this tract. Summers are very hot and the mercury may soar to over 38°C. The SW monsoons cause heavy and widespread rains from July to September. The total annual rainfall is over 200 cms per year with over 90 per cent of it being received from the SW monsoons. Winters are cold and frosty. Rains may occasionally occur in January and February.

The Siwalik sal forest is best developed on the northern slopes of the Siwalik hills in the Corbett National Park and Ramnagar areas. They extend to moist depressions on the south facing slopes also, particularly in the Haldwani area:

The main species occurring in the moist Siwalik sal forests are:

I.	*Adina cordifolia*	*Anogeissus latifolia*
	Bridelia retusa	*Garuga pinnata*
	Lagerstroemia parviflora	*Shorea robusta*
	Terminalia bellerica	*Terminalia tomentosa*
II.	*Aegle marmelos*	*Butea monosperma*
	Cassia fistula	*Gardenia turgida*
	Mallotus philippensis	*Miliusa velutina*
	Randia dumetorum	*Zizyphus xylopyrus*
III.	*Clerodendron viscosum*	*Murraya koengii*
IV.	*Chrysopogon fulvus*	*Heteropogon contortus*
	Themeda sp.	

b) *Terai-bhabar sal forest*: This is another sub-type of the moist sal forest found in the terai and bhabar tract of Kumaun. The soil consists of thick alluvium and boulders. This tract experiences very heavy rainfall in the monsoon season. Summers are extremely hot and the temperature may soar to over 40°C. Winters are mild but frosty.

The main species found in this forest are listed below:

Adina cordifolia	*Anogeissus latifolia*
Bombax ceiba	*Dalbergia sissoo*
Shorea robusta	*Teminalia bellerica*
Terminalia tomentosa	*Zizyphus jujuba*

c) *Dun sal forest*: This is a dense sal forest found in the dun valleys lying between the Siwalik hills and the lower Himalaya. These forests are best developed in the Patli dun valley of Corbett National Park. The forests are fairly dense due to the relatively more moist conditions found in the dun valley.

A typical moist sub-tropical climate is experienced in this tract. The main species forming a part of the dun sal forest are:

I.	*Adina cordifolia*	*Kydia calycina*
	Shorea robusta	*Terminalia bellerica*
	Terminalia tomentosa	
II.	*Mallotus philippensis*	*Miliusa velutina*
	Ougeinia oojeinensis	*Schleichera oleosa*
III.	*Ordisia solanacea*	*Clerodendron viscosum*

d) *Lower Himalayan sal forest*: This is another important sub-type of the moist sal forest. It is found on the lower slopes of the Himalayan foothills. The sal forest extends inland along river valleys, e.g. the Ramganga and Kosi valleys. Well developed in the foothills to the north of Ramnagar.

A typical moist, sub-tropical climate occurs in this tract. The soil is deep and rich in organic matter. The main species found in this forest are:

Albizzia lebbeck	*Anogeissus latifolia*
Shorea robusta	*Terminalia bellerica*
Terminalia tomentosa	

Dry Sal Forest

As the name suggests, the dry sal forest is found under drier conditions usually in the following tracts of Kumaun: (a) dry south facing slopes of the Siwalik hills, and (b) exposed and dry lower Himalayan slopes.

Sal is the dominant species. However, this forest is not as dry as the moist sal forest discussed in the previous text. Biotic pressure is very heavy and hence regeneration is lacking.

Sub-tropical to tropical climatic conditions are experienced in the tract in which this forest is found. Summers are very hot and the

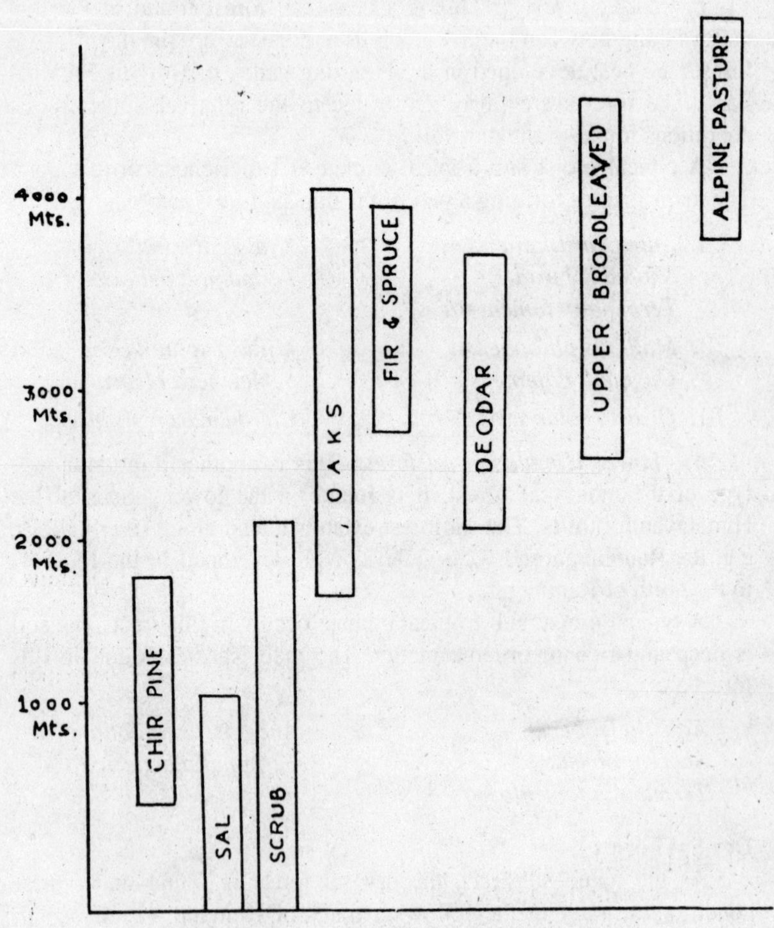

ALTITUDINAL DISTRIBUTION OF FORESTS OF THE KUMAUN HIMALAYA

mercury may soar to over 40°C. The SW monsoons cause heavy and widespread rains. Bulk of the total annual precipitation is recieved in this season. There is forest in winter which leads to the dying of new shoots.

The main species found in this forest are:

I.	*Acacia catechu*	*Anogeissus pendula*
	Dalbergia sissoo	*Diospyros tomentosa*
	Schleichera oleosa	*Shorea robusta*
	Terminalia sp.	
II.	*Cassia fistula*	*Emblica officinalis*
III.	*Colebrookia oppositifolia*	*Dodonea viscosa*
IV.	*Chrysopogon fulvus*	*Eulaliopsis binata*

Lower Mixed Deciduous Forest

This forest type has also been referred to as the northern dry mixed deciduous forest, dry mixed deciduous forest and dry deciduous forest. In Kumaun this forest type is found in the following tracts: (a) on the drier parts of the Siwalik hills, and (b) on the hot and exposed slopes of the lower Himalayan foothills.

The forest is open and is comprised of trees of different deciduous species which are usually of a poor quality. In summer, the maximum temperature rises to more than 38°C. There is adverse dessicating influence of the hot and dry winds from the plains of northern India. The SW monsoons bring relief from the scorching summer heat in early July. They account for bulk of the total annual precipitation which varies from 150 to 200 cms. Winters are mild and frost is common in this season.

The soil is deficient in humus, shallow and in most tracts severely eroded. Biotic pressure is very heavy in the form of grazing and lopping. Regeneration of the tree species is lacking.

The main species found in this forest type are listed below:

I.	*Acacia catechu*	*Aegle marmelos*
	Anogeissus latifolia	*Bombax ceiba*
	Cedrela toona	*Dalbergia sissoo*
	Feronia limonia	*Lannea coromandelica*
II.	*Dendrocalamus strictus*	*Mallotus philippensis*
III.	*Adhatoda vasica*	*Woodfordia fruticosa*
IV.	*Acacia caesia*	*Bauhinia vahlii*

Dry Riverine Forest

This forest type has also been referred to as the khair-sissoo forest. It is essentially a dry deciduous forest found in the following tracts of southern Kumaun: (a) along dry river and stream beds, (b) on freshly laid down alluvium, and (c) on lower hot and exposed slopes.

The dry riverine forest is dominated by khair (*Acacia catechu*) and sisham (*Dalbergia sissoo*) which occur in almost equal proportions. The forest has an open canopy which is usually fairly complete. Individual trees of sisham may attain a height of 15 mts or more. Pure patches of khair may be found on extremely rocky outcrops or on almost pure sandy soils. A fairly dense deciduous undergrowth of shrubs is found in relatively older forests. However, even under dense stands, a large part of the soil remains exposed and is liable to be carried away.

A typical sub-tropical climate prevails in the tract in which this forest type occurs. In summer the temperature may soar to over 40°C. Bulk of the total annual precipitation is recieved from the SW monsoons though winter rains are also fairly heavy. In winter the temperature drops down to about 3°C. Frost commonly occurs during this season. The upper layers of the soil under this forest is highly porous and almost entirely devoid of humus. Water supply is available at moderate depths.

The following three variations may be seen in this forest:

1. Mixed khair and sisham forest with an almost mixed proportion of both.
2. Pure khair forest with occasional trees of sisham.
3. Pure sisham forest with occasional trees of sisham.

These variations may be found in the dry riverine forest based on local site factors such as soil, climate and biotic influences.

The main species occurring in this forest type are listed below:

I. *Acacia catechu* — *Dalbergia sissoo*
Tamarix dioica — *Tiliacora acuminata*

II. *Acacia farnesiana* — *Albizzia lebbeck*
Ehretia laevis — *Holoptelea integrifolia*

III. *Adhatoda vasica* — *Calotropis procera*
Helecteres isora — *Murraya koengii*

IV. *Cassia tora* — *Chrysopogon fulvus*
Erianthus munja — *Saccharum munja*

Dry Bamboo Brakes

These are bamboo which occur in sporadic patches or as thickets within the dry deciduous forests in the following tracts of southern Kumaun: (a) exposed slopes of the Siwaliks, and (b) dry and exposed slopes of the lower Himalaya.

The bamboo brakes consist of clumps of *Dendrocalamus strictus.*

Moist Mixed Deciduous Forest

This forest type has also been referred to as the tropical moist mixed riverine forest. It is found on the moister tracts of the Siwalik hills, dun valleys and the foothills of the lower Himalaya. They may also be associated with sal, khair, sisham and even chir pine forests occurring in the foothills of Kumaun.

The moist mixed deciduous forest is fairly dense. There are a number of dominant species that are intimately mixed together. Individual trees may attain a height of upto 30 mts. The middle canopy is also well developed. It includes a number of evergreen species. There is a heavy growth of climbers.

Typical moist sub-tropical climatic conditions are experienced in this tract. The SW monsoons cause heavy and widespread rains. The main species occurring in this forest type are:

I.	*Adina cordifolia*	*Albizzia procera*
	Bauhinia retusa	*Bombax ceiba*
	Cedrela toona	*Dalbergia sissoo*
	Ficus sp.	*Lannea coromandelica*
	Madhuca indica	*Shorea robusta*
	Terminalia bellerica	*Terminalia tomentosa*
II.	*Bauhinia vahlii*	*Cassia fistula*
	Garuga pinnata	*Kydia calycina*
III.	*Adhatoda vasica*	*Helicteres isora*
IV.	*Barleria strigosa*	*Costus speciosus*

Lower Scrub

These are scrub forests that are found on the dry exposed and degraded slopes of the following tracts of southern Kumaun: (a) the dun valleys and Siwalik hills, and (b) foothills of the lower Himalaya.

This forest type may further be sub-divided into the following sub-types:

a) *Dry scrub forest*: This sub-type may further be of the following kinds:

1. A scrub containing a fair proportion of trees which may attain a height of upto 20 mts.

2. A scrub containing a fair proportion of tree growth and a relatively open undergrowth of brushwood species.

3. A scrub containing occasional/sporadic trees and an open undergrowth of brushwood species.

The dry scrub forest is made up of the following species:

Acacia catechu
Albizzia lebbeck
Anogeissus latifolia
Diospyros tomentosa
Lannea grandis

b) *Dry deciduous scrub*: This is a dry deciduous scrub in which shrubs having a height between 3 to 6 mts. dominate. Occasional small trees and clumps of bamboos may also occur. This scrub is under heavy biotic pressure.

The dry deciduous scrub consists of the following species:

Acacia catechu
Aegle marmelos
Carissa opaca
Dodonea viscosa
Gardenia turgida
Zizyphus sp.

Chir Pine Forest

As the name suggests, this forest type is made up almost entirely of a pure crop of chir pine (*Pinus roxburghii*) which forms the top canopy. Individual chir pine trees may attain a height of upto 40 mts.

Typical sub-tropical climatic conditions are experienced in the tract in which this forest occurs. The summer temperature ranges from 20°C to 35°C while in winter the mercury may drop to near the freezing point. The SW monsoons cause heavy and widespread rains. They account for bulk of the total annual precipitation which may be over 200 cms. Chir pine forests are affected by fires in the summer season every year.

The soil is sandy and poor in humus content. This forest attains best growth over quartzitic rocks. The forest floor is covered by a good growth of perennial and annual grasses.

The chir pine forest is comprised of the two sub-types discussed below:

a) *Siwalik chir pine forest*: Also known as the lower chir pine forest this forest type is well developed on the tops of the Siwalik hills in southern Kumaun, usually above an elevation of 800 mts.

Sub-tropical climatic conditions prevail in this tract. Summers are hot and dry and the average maximum temperature may be over 38°C. Dusty hot winds from the Ganga plains lash the tops of the Siwalik hills. The rains set in by late June. They cause heavy and widespread rains till mid-September. Autumn lasts till November after which the winter season begins. Frost is common during the cold season. Winter rains are also recieved in this tract.

The Siwalik chir pine forest is a more or less pure association of chir pine. The forest is relatively open and the site quality is low. The main constituents of this forest are:

I. *Pinus roxburghii*

II. *Acacia catechu*, *Adina cordifolia*, *Albizzia lebbeck*, *Dalbergia sissoo*, *Emblica officinalis*, *Mallotus philippensis*, *Pyrus pashia*, *Terminalia chebula*

III. *Carissa opaca*, *Crataegus crenulata*, *Dodonea viscosa*, *Myrsine africana*

The following variations may occur in the composition of the Siwalik chir pine forest:

1. Chir pine-*Carissa*-grass with *Terminalia chebula* and *Lannea*
2. Chir pine-*Carissa*-Indigofera-grass with abundant regeneration of pine
3. Chir pine-*Ban* oak-*Berberis* and *Myrsine* with grass
4. Chir pine-Khair-*Mallotus-Lannea* with grass
5. Chir pine-*Wendlandia-Dodonea* and *Carissa*
6. Chir pine-*Carissa-Flacourtia* and grass.

b) *Himalayan chir pine forest*: This has also been referred to as the upper chir pine forest. It is more or less a pure forest of chir pine occurring upto an elevation of about 1800 mts in many parts of Kumaun. This forest is found in the following physiographic tracts: (a) lower and middle Himalaya, and (b) foothills of the main Himalaya.

Himalayan chir pine forests are best developed in Bhowali, Ranikhet and Almora areas. These forests are abundant over rocks rich in quartz, e.g. quartzite, sandstone and granite and on soils rich in sand.

Summers are hot and dry. Bulk of the total annual precipitation is received from the SW monsoons though winter rains are also fairly heavy. Snowfall may occur at higher elevations but it usually does not remain on the ground for long. The average maximum temperature of this tract ranges from 38°C to 40°C.

These forests are under heavy biotic pressure. Chir trees are extensively used locally for the construction of houses and are also tapped for resin. This has led to the degradation of some of these forests particularly near human habitation. Frequent fires also occur in summer. They damage or hamper regeneration. Grazing is also harmful for the growth of seedlings.

The upper or Himalayan chir pine forest is denser than its counterparts found on the tops of the Siwalik hills. The site quality is excellent in many areas of Almora district. Individual trees attain a height of more than 40 mts.

The main species occurring in this forest are listed below:

I. *Pinus roxburghii*

II. *Coriaria nepalensis* — *Crataegus crenulata*
Ficus roxburghii — *Lyonia ovalifolia*
Myrica sapida — *Quercus leucotricophora*
Rhododendron arboreum — *Syzygium cumini*

III. *Glochidion velutinum* — *Rubus ellipticus*

IV. *Artemisia vulgaris* — *Eulalia mollis*

V. *Rosa moschata*

Ban Oak Forest

This is a broadleaved forest in which ban oak (*Quercus leucotricophora*) is the main species. These forests occur between an elevation of 1700 and 2200 mts in the lower temperate belt of Kumaun, viz. in parts of lower, middle and higher Himalaya. Well developed in Rani- khet and Nainital areas.

The mean annual temperature in this tract ranges from 18°C to 20°C. The total annual rainfall is from 150 to 200 cms. There is heavy biotic pressure on these forests in the form of lopping for fuelwood and removal of fodder. As a result of this, many oak forests near human habitation have been severely degraded.

The ban oak forests are best developed on sheltered aspects and moist shady depressions. The main species forming a part of this forest are:

I.	*Carpinus viminea*	*Lyonia ovalifolia*
	Pinus roxburghii	*Quercus leucotricophora*
	Rhododendron arboreum	*Toona serrata*
II.	*Euonymus pendulus*	*Ilex dipyrena*
	Lindera pulchersima	*Rhus parviflora*
III.	*Berberis lycium*	*Deutzia staminea*

Moru Oak Forest

This is a dense temperate broadleaved oak forest found between an elevation of 2200 and 2400 mts. It is dominated by moru oak (*Quercus himalayana*) with individual trees attaining a height of over 20 mts. The forest is fairly dense and the upper, middle and lower storeys are well developed, particularly on moister aspects.

In Kumaun, the moru oak forest is best developed on the middle and main Himalayan ranges, e.g. in Kausani, Binsar and Pithoragarh areas.

This forest type occurs immediately above the ban oak forest described in the above text. The climatic conditions in this tract are cooler. Summers are mild and mercury does not usually rise to more than 24°C. Winters are very cold and severe. There is heavy snowfall and the snow may lie on the ground for several days at a stretch in this season. These forests too are under heavy biotic pressure, particularly near human habitation. The soil under this forest is very rich in humus.

The main species forming a part of this forest type are:

I.	*Acer caesium*	*Betula alnoides*
	Quercus himalayana	*Quercus leucotricophora*
	Quercus semecarpifolia	
II.	*Eurya acuminata*	*Fraxinus micrantha*
	Litsea umbrosa	*Pyrus lanata*
	Rhododendron arboreum	*Taxus baccata*
III.	*Daphne canabina*	*Indigofera gerardiana*
IV.	*Clematis montana*	*Hedera nepalensis*

Kharsu Oak Forest

This forest type occupies the highest elevation amongst the three main oak forest in Kuamun. It occurs between an elevation of 2500 and 3500 mts. As the name suggests this forest is dominated by Kharsu oak (*Quercus semecarpifolia*).

The tract in which this forest occurs experiences a typical temperate climate. Summers are mild and the maximum temperature usually does not exceed 23°C. Winters are severe and sub-zero temperatures are common. There is heavy snowfall in this season and the snow may lie on the ground for several days at a stretch. Hailstorms may occur in the short spring season that follows the long winter. Kharsu oak forest are under heavy biotic pressure near human settlements.

The canopy is fairly dense and the trees attain a height of upto 20 mts or more. The upper storey may also consist of other broadleaved and coniferous species though their proportion as compared to kharsu oak is quite low. These forests are very dense in moist shady depressions.

In Kumaun, kharsu oak forests are well developed in parts of Almora and Pithoragarh districts, usually on the middle and main Himalayan slopes.

The main species forming a part of the kharsu oak forest are:

I. *Acer caesium* — *Abies pindrow*
Betula alnoides — *Picea smithiana*
Pyrus lanata — *Quercus himalayana*
Quercus semecarpifolia

II. *Betula utilis* — *Prunus padus*
Rhododendron arboreum — *Rhododendron barbatum*
Sorbus foliolosa — *Taxus baccata*

III. *Ribes glaciale* — *Rosa macrophylla*
Sarcococca saligna — *Skimmia larureola*

IV. *Ainsliea aptera* — *Polygonum speciosum*

V. *Clematis montana* — *Vitis* sp.

Moist Deodar Forest

This is a more or less pure forest of deodar found between an elevation of 2000 and 2500 mts in the following tracts: (a) upper reaches of the lower Himalaya, and (b) middle and main Himalayan ranges.

Deodar forests are well developed in parts of Almora and Pithoragarh districts.

The canopy is more or less closed and individual trees may attain a height of over 40 mts. Deodar trees have long clear boles and whorled branches, thus presenting a majestic appearance. Other conifers like fir, spruce and blue pine may also form a part of the top canopy but deodar is by far the most dominant species. Broadleaved trees occur along nalas and in moist shady depressions.

Summers are mild and the mercury rarely rises to more than 25°C. The SW monsoons cause heavy and widespread rains in the entire tract in which the deodar forest occurs. Winters are long and severe and the temperatures drop down to below the freezing point. Heavy snowfall occurs at frequent intervals in the winter season. It aids in the establishment of regeneration.

The main species occurring in this forest are listed below:

I. *Abies pindrow* | *Cedrus deodara*
Picea smithiana | *Pinus roxburghii*
Pinus wallichiana

II. *Quercus himalayana* | *Quercus leucotricophora*
Quercus semecarpifolia | *Rhododendron arboreum*
Taxus baccata

III. *Ribes glaciale* | *Smilax vaginata*

IV. *Fragaria vesca* | *Polygonum speciosum*

Temperate or Upper Moist Mixed Deciduous Forest

This forest has also been referred to as the upper moist mixed deciduous forest or moist temperate deciduous forest. As the name suggests, this is mixed deciduous high forest found between an elevation of 2000 and 3000 mts in the following parts of Kumaun: (a) upper reaches of the lower Himalaya, and (b) middle and main Himalayan ranges.

The upper canopy is made up of a number of broadleaved species and occasional coniferous trees. Individual trees may attain a height of more than 30 mts. They show best development along rivers and streams and in moist shady depressions of north facing slopes.

Temperate climatic conditions prevail in the tract in which this forest occurs. In summers the temperature is not usually more than 25°C. The SW monsoons cause heavy and widespread rains. A significant part of the total annual precipitation is in the form of snow which may cover the ground for several days at a stretch in the winter season. In the cold season temperatures fall below the freezing point.

The soil under these forests is fertile and rich in humus. The temperate moist mixed deciduous forests are under intense biotic pressure particularly near human habitations. These forests are best developed in the upper tract of Almora district and in the upper and middle catchment of the Kali river.

The following are the main species forming a part of this forest:

I.	*Acer caesium*	*Acer pictum*
	Acer villosum	*Aesculus indica*
	Betula alnoides	*Celtis australis*
	Fraxinus micrantha	*Juglans regia*
	Prunus cornuta	*Pyrus lanata*
II.	*Corylus colurna*	*Euonymus tingens*
	Rhus sp.	*Taxus baccata*
III.	*Spiraea aruncus*	*Spiraea cestita*
IV.	*Hedera nepalensis*	*Vitis lanata*

Upper Scrub

The upper scrub forest has also been termed as the temperate secondary scrub. These are low scrub forests made up of small sized trees and shrubs. Thorny species are dominant on hot, dry exposed slopes. In the rainy season, shrubby growth is dense and abundant. This scrub forest is found on exposed slopes of the temperate zone of Kumaun.

The tract in which this forest occurs experiences mean maximum temperature of about 26°C. Summers are warm and dry. The total annual precipitation is both in the form of rain and snow. Winters are very severe with temperatures falling to levels below the freezing point. The spring season is short.

This is a secondary forest that has developed in localities that are under a heavy biotic pressure. The main species occurring in this forest are:

Berberis lycium	*Cotoneaster baccilaris*
Carissa opaca	*Dodonea viscosa*
Indigofera gerardiana	*Olea cuspidata*
Prinsepia utilis	*Pyrus pashia*
Rosa macrophylla	*Spiracea canescens*

Temperate Mixed Coniferous Forest

This forest types has also been referred to as the western mixed coniferous forest. It is a mixed forest consisting of attractive coniferous trees, viz. deodar, blue pine, fir and spruce which occur in varying proportions. The altitudinal range of this forest type is from 2400 to 3000 mts or even higher.

In Kumaun, the temperate mixed coniferous forest is found in the middle and main Himalayan ranges, particularly in the upper valley of the Kali river.

Even though this forest is dominated by conifers, there may also occur broadleaved trees, particularly in moist, shady depressions. Patches of grassy meadows are also found sporadically within the temperate mixed coniferous forest.

A typical temperate climate is experienced in the tract in which this forest is found. Snowfall accounts for a substantial part of the total annual precipitation. The first snowfall of the winter season may occur as early as in late November. In winter the mercury remains below the freezing point for long periods. Summers are mild and the average maximum temperature does not usually exceed 28°C.

The main species which form a part of the temperate mixed coniferous forests are:

I.	*Abies pindrow*	*Cedrus deodara*
	Picea smithiana	*Pinus wallichiana*
II.	*Acer acuminatum*	*Acer pictum*
	Enonymus lacerus	*Betula utilis*
	Quercus sp.	*Taxus baccata*
III.	*Deutzia corymbosa*	*Ribes rubrum*
	Skimmia sp.	*Viburnum nervosum*
IV.	*Adinatum venustum*	*Aspidium aculeatum*
V.	*Clematis barbellata*	*Schizandra grandiflora*

Oak Fir Forest

This forest type has also been referred to as the upper oak fir forest. The oak-fir forest is found above an elevation of 2600 mts and may extend upto the timber line. It occurs in the main Himalayan ranges of Kumaun.

Individual trees attain a height of more than 25 mts. Both coniferous and broadleaved species are intricately mixed together. The proportion of the latter is more in moist localities. Sub-arctic to arctic climatic conditions prevail in this tract. Summers are warm and dry. The SW monsoons cause rain and hailstorms from late June to the middle of September. Autumn is short and the winter season begins as early as mid-November. Very heavy snowfall occurs in the cold season. It lies on the ground for several weeks at a stretch. The mercury remains below the freezing point in this season.

The soil is rich in humus, well areated and moderately deep. A thick layer of undecomposed organic matter covers the forest floor. The main species which constitute this forest type are:

I. *Abies pindrow* — *Cedrus deodara*
Picea smithiana — *Pinus wallichiana*

II. *Acer* sp. — *Betula alnoides*
Betula utilis — *Corylus colurna*
Pyrus lanata — *Quercus semecarpifolia*

III. *Berberis* sp. — *Indigofera gerardiana*
Ribes glaciale — *Rosa macrophylla*
Rosa sericea — *Viburnum nervosum*

IV. *Fragaria vesca* — *Podophyllum hexandrum*

V. *Clematis montana* — *Schizandra grandiflora*
Vitis semicordata

Cypress Forest

This is by and large an open forest consisting of sporadic trees of cypress. It is found on steep rocky tracts in the lower, middle and main Himalayan ranges, between an elevation of 1800 mts and 2800 mts. Cypress forests attain best development over limestone rocks and calcerous soils.

Typical temperate climatic conditions prevail in the tract in which this forest is found. Bulk of the total annual precipitation is received from the SW monsoons though rain and snow are also received in winter. Summers are fairly warm and dry. The soil is poor in humus, exposed, shallow and prone to erosion.

The main species forming a part of this forest types are:

I. *Cedrus deodara* — *Cupressus torulosa*

II. *Aesculus indica* — *Juglans regia*
Pyrus pashia

III. *Abelia triflora* — *Cotoneaster bacillaris*
Prinsepia utilis — *Rosa* sp.

Alder Forest

This is an almost pure forest of alder (*Alnus nitida*) with occasional trees of other species. The alder forest is found along the banks of rivers and streams upto an elevation of about 3000 mts in the middle and main Himalayan valleys.

Temperate climatic conditions are experienced in this tract. The soil consists of recently laid down alluvium. The main species forming a part of this forest are:

I. *Alnus nitida* — *Celtis australis*
Populus ciliata — *Populus deltoides*
Ulmus villosa

II. *Crataegus crenulata* — *Sarcococca*

Blue Pine Forest

This forest type has also been referred to as the riverine or low level blue pine forest. As the name suggests, this is virtually a pure forest of blue pine (*Pinus wallichiana*) which is found between an elevation of 2100 and 2500 mts in the middle and main Himalayan ranges of Kumaun. Occasional trees of other coniferous and broadleaved species are associated with blue pine. The latter may occur in moist shady depressions.

Temperate to sub-arctic climatic conditions are exprienced in the tract in which this forest type occurs. Rainfall occurs both during the monsoon season and in winter. Winters are very cold and severe while summers are mild.

The main species forming a part of this forest type are:

I. *Abies pindrow* — *Cedrus deodara*
Picea smithiana — *Pinus roxburghii*
Pinus wallichiana

II. *Aesculus indica* — *Juglans regia*
Quercus sp. — *Rhododendron arboreum*

III. *Indigofera gerardiana* — *Prinsepia utilis*

Dry Temperate Forest

These forests are found in the dry tracts of the higher and trans-Himalaya which lie in the rain-shadow of the main Himalayan mountain wall. Very low rainfall occurs in this area and bulk of the total annual precipitation is in the form of snow. Various forms and subtypes of forests found in the upper dry tracts of Kumaun are:

a) *Mixed forest*: This is an open forest comprising of both broadleaved and coniferous species. The altitudinal range is from 2000 to 2400 mts. Summers are mild but winters are very severe in the tract in which this forest occurs. Rainfall is very low.

b) *Dry deodar forest*: This is a more or less pure forest of deodar found in the dry zone of the higher Himalaya. Individuals of other broadleaved and coniferous species may also occur between the pure patches of deodar.

c) *Dry blue pine forest*: This is a more or less pure forest of blue pine found in the inner dry valleys of the higher Himalaya between an elevation of 3000 and 3600 mts. In this tract too, rainfall is very low.

d) *Dry juniper scrub*: This is an evergreen, open forest comprised of dwarfed and stunted junipers. It is found between an elevation of 2700 and 4300 mts in the inner dry valleys of the main Himalaya.

Sub-Alpine Forest

The sub-alpine forest represents the highest form of tree forest in Kumaun. It extends from elevations of about 3000 mts. to the timber line in the main Himalayan mountain wall of northern Kumaun.

This forest type may further be divided into the following subtypes:

a) *Upper fir forest*: This forest is found between an elevation of 2900 and 3000 mts in the higher Himalayan ranges. Arctic climatic conditions are experienced. Winters are very cold and a thick blanket of snow covers the ground for several weeks at a stretch in this season. There is a thick layer of humus on the forest floor.

The main species which constitute this forest are:

I. *Abies spectabilis* — *Picea smithiana*
Pinus wallichiana

II. *Prunus padus* — *Taxus baccata*

b) *Birch-fir forest*: This forest is found above an elevation of 3000 mts. The trees are stunted with a low spreading crown. They tend to attain a bushy growth. The main species occurring in this forest are:

I. *Abies spectabilis*

II. *Betula utilis* — *Quercus semecarpifolia*
Rhododendron sp. — *Sorbus foliolosa*

III. *Cotoneaster acuminata* — *Ribes glaciale*
Ribes rubrum — *Rubus nivens*
Smilax vaginata

c) *Birch-rhododendron scrub forest*: This is a low evergreen forest dominated by birch and rhododendron. The trees are stunted and bushy. Their stems may be slightly curved due to the pressure of snow that covers the ground for considerable periods of time during the winter months. The mercury drops to level below the freezing point in winter. Summers are mild and wet.

The main species forming a part of this forest are:

I. *Betula utilis* — *Rhododendron* sp.
Sorbus foliolosa

II. *Berberis* sp. — *Lonicera parviflora*

III. *Polygonum* sp. — *Viburnum nervosum*

Alpine Forest

As the name suggests, this forest type is found just below the line of perpetual snow in the main Himalayan ranges and inner dry valleys of Kumaun; usually occurring above an elevation of about 3600 mts.

This forest may further be sub-divided into:

a) *Moist alpine scrub*: This is a scrub or meadow occurring near the snowline above an elevation of about 3300 mts. It consists of low deciduous shrubs which rarely grow to more than 2 mts in height. Summers are mild and warm in this tract while winters are extremely cold. Thus scrub is subjected to heavy grazing during the summer season. The main species are:

I & II. *Betula utilis* — *Rhododendron* sp.

III. *Berberis* sp. — *Lonicera* sp.
Syringa emodi

IV. *Aconitum* sp. — *Trollis* sp.

b) *Dwarf rhododendron scrub*: This is a scrub or alpine meadow with sporadic growth of dwarfed rhododendron occurring near the snowline. It succeeds the sub-alpine forest and merges with the line of perpetual snow.

The main species occurring in this scrub or meadow are:

Lonicera obovata — *Rhododendron anthopogon*
Rhododendron campanulatum — *Sorbus foliolosa*

c) *Dry scrub*: This scrub vegetation occurs in the inner dry valleys lying in the rainshadow of the main Himalayan range. There is very little rainfall in this tract. The soil depth varies from shallow to very thin and it is very poor in humus.

The dry scrub forest is very well developed along streams formed by snow melt waters. The principal species forming a part of this scrub are:

Arenaria sp.
Artemisia maritima
Artemisia sacrorum
Caragana sp.
Draba graciallima
Eurotia ceratoides
Juniperus communis
Juniperus wallichiana
Potentialla fruticosa
Primula sp.
Saxifera imbricata
Sedum crassipes

d) *Dwarf juniper scrub*: This is another scrub forest found in the inner dry valleys, e. g. upper Darma valley. Winters are very cold and snow covers the mountain slopes for 5 to 6 months in a year. This scrub is dominated by junipers.

7

Wildlife and Sanctuaries

Hundreds of species of wild animals ranging from elephants and tigers to small lizards are found in Kumaun. This region represents a transition zone of the western and central Himalayan fauna. The fauna of Kumaun is akin to that of the Indo-Ganga plain; Tibet, western and central Himalaya. In this chapter important animals found in this region have been dealt with. The national parks and sanctuaries of Kumaun have also been described.

MAMMALS

Mammals are a group of animals who suckle their young. Their body temperature remains more or less the same irrespective of the atmospheric temperature. In Kumaun, mammals are found right from the foothill zone along the Ganga plain to the snowline.

Amongst the larger mammals of Kumaun are the elephant and tiger which are found in the lower hills. In the past the Indian one-horned rhinoceros was also found in the foothills. However, today it is found in Nepal (east of Kumaun). Other larger mammals of Kumaun include the leopard, snow leopard (occasional) and brown and black bears.

The medium sized mammals found in Kumaun include the sambhar, cheetal, musk deer, hog deer, ghoral, bharal, hyaena, jackal, wolf and wild boar.

The smaller mammals include bats, primates, rats, rabbits and porcupines.

Different species of mammals found in Kumaun have been discussed in the following text (after Prater, 1972 and Negi, 1992).

Primates

1. *Rhesus macaque* (*Macaca mulatta*): This is the common monkey found in and around human settlements in the lower and middle Himalaya of Kumaun. In summer, troops of this species may ascend to elevations of over 2500 mts in search of food. They are good swimmers and are able to cross even rapid torrents. The rhesus macaque eats fruits, leaves, succulent shoots and roots, spiders and insects.

2. *Assamese macaque* (*Macaca assamensis*): This species is not very common in Kumaun. It is found in the lower Himalaya between 800 and 1200 mts. They live in dense forests and avoid contact with human beings. Their food consists of fruits, seeds, leaves, roots, shoots and insects.

3. *Common langur or Hanuman monkey* (*Presbytis entellus*): The common langur is found upto an elevation of about 3500 mts in different parts of Kumaun. They move in troops of varying sizes and ascend to the sub-alpine forests in search of food in summer. Their food consists of fruits, flowers, leaves, shoots, succulent roots and even crops.

Cat family

1. *Tiger* (*Panthera tigris*): The tiger is found in the dense forests of the lower hills of Kumaun. In the Corbett National Park area it may ascend to elevations of about 1500 mts. The tiger is a solitary animal and may cover a large distance with the span of 24 hours. It preys on deers, antelopes, wild boars, primates and even domestic animals.

2. *Leopard or Panther* (*Panthera pardus*): The leopard is found all over Kumaun from the foothills to tracts near the timber line. It lives both in the forests and in open rocky areas. The leopard will kill and eat anything it can catch hold of, viz. deers, antelopes, wild boars, porcupines, monkeys, birds, domestic cattle and even poultry.

3. *Snow leopard or Ounce* (*Panthera unica*): This is an elusive animal found at very high elevations near the snowline. Its existence in Kumaun is very rare and occasional. The snow leopard lives in caves in the stunted forests lying near the snowline. It stalks the alpine pastures in search of prey during the summer months.

4. *Leopard cat* (*Felis bengalensis*): The leopard cat is a little larger than the domestic cat and has larger legs. It is found in the lower Himalaya both in forested areas and around human settlements. It preys upon smaller mammals, birds, domestic animals and poultry.

5. *Fishing cat* (*Felis viverrina*): This animal has a size more than that of a domestic cat with short limbs and stout built. It is found in the foothills of Kumaun, usually upto an elevation of about 1500 mts. It lives in or near thick forests or in scrublands usually around grassy swamps, rivers, streams and lakes. The fishing cat preys on any animal it can kill including fishes and fresh water molluscs. It does not usually enter water to catch its prey and prefers to fish by crouching on a rock or overhanging bank.

Civets

1. *Small Indian civet* (*Viverricula indica*): Found only in the foothills usually preferring tall grasslands. It searches its food on the ground.

2. *Common palm civet or Toddy cat* (*Paradoxurus hermaphroditus*): A black or dark brown civet bearing long and coarse hair. Found in the foothills. It lives in forest areas and spends the day amongst the branches or in a hole in the tree-trunk. It hunts for food at night.

3. *Himalayan palm civet* (*Paguma larvata*): Found in the foothills of Kumaun. It lives in the holes of trees or on the branches and hunts for food either on tree-tops or on the ground. The Himalayan palm civet eats fruits, birds and rodents.

Mongooses

1. *Common mongoose* (*Herpestus edwardsi*): This is a small, twany, yellowish grey animal with no stripes on the side of its neck. Found in open areas, scrublands and in and around cultivated areas of the foothills.

2. *Small Indian mongoose* (*Herpestes auropunctatus*): This animal is smaller in size than the common mongoose. It has a shorter tail and bears soft silky fur. Found in the foothills of Kumaun. It searches for food both during the day and at night.

Hyaenas

1. *Striped hyaena* (*Hyaena hyaena*): This is a dog-like animal found in the open areas and scrublands of the lower hills. They either kill small animals and birds or eat the remains of a tiger or leopard kill. Hyaenas are usually nocturnal in habit, preferring to remain hidden during the day.

Dog Family

1. *Jackal* (*Canis aureus*): This is a dog-like animal found in different parts of Kumaun usually between an elevation of 1200 and 2100 mts. However, jackals may ascend to elevations of over 3000 mts in search of food. Jackals live in dense forests, open areas, grasslands, scrublands and in the vicinity of lakes, streams and rivers. This animal usually comes out in search of food at dusk and return to their shelters at dawn. They hunt in ones, twos or in small packs. Jackals may also kill poultry or small domestic cattle.

2. *Indian fox* (*Vulpes bengalensis*): This is a slender limbed animal with a grey fur whose shade varies from season to season. Found in the open areas and scrublands of the foothills of Kumaun. The fox eats fruits, small animals and birds.

Bear Family

1. *Himalayan black bear* (*Selenarctos thibetanus*): This is a relatively large bear with a thick fur. Found in many thickly forested tracts of Kumaun. They are altitudinal migrants and move to elevations of over 3500 mts near the snowline in search of food during the summer season. In winter they return to the shelter of the valleys and may descend to about 1500 mts. The black bear spends the day sleeping in a cave and comes out to search for food about dusk. It is both a herbivore and carnivore.

2. *Brown bear* (*Ursus arctos*): This is an animal of heavy built having a brown coat whose shade varies from season to season. The fur becomes thick just before the onset of winter. It is found in the upper tracts of Kumaun, usually in open rocky areas above or near the tree-line. They descend to lower elevations in winter. The brown bear is both a herbivore and carnivore.

Weasel Family

1. *Common otter* (*Lutra lutra*): Found in and around cold hill and mountain lakes, streams and rivers of Kumaun. It lives amongst rocks and boulders and in the hollows of trees near waterbodies. They may ascend to elevations of over 3000 mts along rivers and streams in search of food during the summer season. This coincides with the upward movement of carp and other fishes for spawning in summer. The main food of common otter consists of fishes, amphibians, crabs and small aquatc birds.

2. *Smooth Indian otter* (*Lutra perspicillata*): A relatively heavily

built animal found in the foothills of Kumaun. It lives around lakes, streams and rivers. They eat fishes, amphibians and birds. This animal is active both on land and in water.

3. *Clawless otter* (*Aonyx cinera*): This species is found in the marshes, streams and rivers of the foothills. Its claws are reduced to small spikes.

4. *Stone marten or Beech marten* (*Martes foina*): A slender animal with long legs. It has a fairly long tail. Found above elevations of over 1600 mts. They inhabit forested areas, rocky tracts and grasslands. The stone marten hunts for food both during the day and at night. Their food consists of anything they can kill, including squirrels, hares, frogs, lizards, snakes, honey, nuts and fruits.

5. *Himalayan yellow throated marten* (*Martes flavigula*): This animal is of a relatively larger size than the stone marten described in the above text. Found between an elevation of 1200 and 2700 mts. The yellow throated marten usually lives and hunts alone. It is both a herbivore and carnivore. They normally avoid areas around human habitation.

6. *Himalayan weasel* (*Mustala sibirica*): This animal is found between an elevation of 1500 and 4800 mts. It lives in open areas, scrublands and forested tracts. The Himalayan weasel may also be found near the snowline in summer. They eat anything they can catch hold of.

7. *Yellow bellied weasel* (*Mustala kathiah*): Found in different forested tracts of Kumaun.

8. *Honey badger or Ratel* (*Mellivora capensis*): This animal has a squat, bear-like body with stumpy, short legs and a small tail. It is found in the relatively drier tracts of the Himalayan foothills. The honey badger eats small mammals, birds, reptiles, amphibians, insects, fruits, roots and honey.

Bats

1. *Flying fox* (*Pteropus giganteus*): A large sized bat found upto an elevation of about 2200 mts. It prefers to remain in forested areas. The flying fox relishes the juice of fruits while the pulp is spat out.

2. *Fulvous fruit bat* (*Rousettus leschenaulti*): This is a medium sized bat having a light brown colour. It is gregarious while roosting. This is occasionally seen in the foothills of Kumaun; not very common.

3. *Great Himalayan leaf-nosed bat* (*Hipposideros arniger*): This

is a large, brown bat found in the lower hills upto an elevation of about 2000 mts. They usually come out in search of food after dusk.

4. *Serotine* (*Eptesicus serotinus*): This is a dark coloured palaearctic bat found in the temperate forests of Kumaun. It hibernates in small groups in winter usually in the hollows of trees and in caves.

5. *Indian pipistrella* (*Pipistrelle coromandra*): This bat is dark brown in colour. It is found in the lower Himalayan forests of Kumaun. The pipistrella hibernates in winter. It may fly almost all through the night in search of food.

6. *Common yellow bat* (*Scotophilus heathi*): A yellowish brown coloured bat found in and around human settlements in the foothills. This animal too hibernates in winter.

Rodent Family

1. *Red flying squirrel* (*Petuarista petuarista albiventer*): This is a bright chestnut coloured squirrel found in different parts of the Kumaun hills. It lives in the hollows of tree trunks or in caves.

2. *Five striped palm squirrel* (*Fundabulus pennanti*): This animal is found in the foothills of Kumaun, particularly in the drier tropical and sub-tropical tracts.

3. *Himalayan striped squirrel* (*Callosciurus macclellandi*): A greyish brown coloured squirrel with black and brown stripes on its back. Found in different parts of Kumaun, usually above an elevation of about 1500 mts. This is a shy and elusive animal.

4. *Himalayan marmot* (*Marmota bobak*): This animal is found between an elevation of 4000 and 5300 mts. It lives in burrows where they hibernate in winter. The marmots emerge from their burrows as the snow begins to melt in spring. They feed on roots, shoots, grasses, leaves, fruits and seeds.

5. *Indian gerbille* (*Tatera indica*): The Indian gerbille inhabits the open areas, scrublands and cultivated tracts of the foothills of Kumaun.

6. *Indian mole rat* (*Bandicota bengalensis*): They live in the vicinity of agricultural fields in different parts of the Himalaya. They dig extensive burrows.

7. *Field mouse* (*Mus boodunga*): Common in fields, pastures and open areas of the lower Himalayan foothills.

8. *Long-tailed tree mouse* (*Vandeleuria oleracea*): This mouse is found in the foothills of Kumaun. It lives both on trees and shrubs. Its

food consists of tender shoots, buds, fruits and leaves. The long tail gives it a better hold while climbing trees.

9. *Common house rat* (*Rattus rattus*): Commonly found in and around human habitation in the foothills, lower and middle Himalaya of Kumaun.

10. *Bandicoot rat* (*Bandicota indica nemorivaga*): Found in the vicinity of human habitation in the foothills, lower and middle Himalaya.

11. *Common house mouse*: Found in and around human habitation in the foothills of Kumaun. The common house mouse is omnivorous and will eat virtually any food it can get hold of.

12. *Indian porcupine* (*Hystrix indica*): The Indian porcupine is commonly found in the forests and open areas of Kumaun, from the foothills to an elevation of about 2400 mts. They prefer rocky hillsides. Its erect quills are used for killing prey and as a mechanism of defence.

Elephant

Indian elephant (*Elephas maximus*): This is the largest animal found in India. In Kumaun, the elephant is found in the dun valleys, Siwalik hills, terai and bhabar tract and the foothills of the lower Himalaya. They usually do not ascend to elevations of over 1000 mts.

The males have large tusks while those of females protude for a few centimetres. Elephant herds are made up of 5 to 50 individuals. They migrate over long distances. Elephants are herbivores. Tuskers or loners may roam individually and are highly dangerous.

Rhinoceros

Great Indian one-horned rhinoceros (*Rhinoceros unicornis*): This animal was once found in the foothills of Kumaun. It is now extinct from this tract.

Wile Oxen, Sheep and Goats

1. *Yak* (*Bos grunniens*): A large, heavily built animal having a slightly drooping head, straight back, high humped shoulders and short, sturdy, fore and hind legs. It has been domesticated in the upper tracts of Pithoragarh. Wild yaks ascend to elevations of over 4200 mts in summers, usually searching for food in the alpine meadows. They return to relatively lower elevations in winter. This animal is able to live under the most difficult climatic and physical conditions found

anywhere in the Himalaya. Wild yak are very rare in Kumaun.

2. *Great Tibetan sheep or Nayan* (*Ovis ammon hodgsoni*): Found occasionally in the upper Darma valley of Kumaun. They live on dry, desolate slopes. The climatic conditions of this tract are very harsh. Summers are dry and hot while very heavy snowfall is received in winter. Vegetation begins to sprout after the snow-melts in late spring.

3. *Bharal*: In appearance, the bharal resembles both a sheep and a goat. Its horns are smooth, rounded and form a curve backwards over the neck. The fur is brownish grey in colour which attains a slaty grey hue in winter and becomes browner in summer. In Kumaun the bharal is found in the desolate tracts of northern Pithoragarh, usually on the slopes of the main Himalayan range where it lives between the timberline and the snowline. They inhabit grasslands and forested areas; feeding on grasses, mosses and dwarf shrubs. Flocks ascend to higher elevations in search of food during the summer season.

4. *Ibex* (*Capra ibex*): A stoutly built animal whose horns are curved backwards over the neck. The male ibex has a beard and a coat of coarse brittle hairs. It develops a thick winter coat which attains a higher shade in summer.

The Ibex is found on the slopes of the main Himalayan range in northern and north-western Kumaun, usually above an elevation of over 3800 mts. A herd may be made up of 30 to 40 individuals. They feed on grass, moss and small shrubs. Herds of ibex move upto higher elevations with the retreating snow in summer to return to the shelter of lower tracts in winter.

5. *Wild goat* (*Capra hircus*): This is a sturdy animal whose horns curve back towards the neck. Its coat is rufous brown in summer and attains a brownish grey shade in winter. Found in northern Kumaun, where it is not very common.

6. *Himalayan thar* (*Hemitragus jemlahicus*): This is a species of wild goat with a heavy body and long robust limbs. Facial hair is short though the body is covered with coarse hairs that is tangled into thick masses.

The Himalayan thar is found in the sub-alpine zone of northern Kumaun between an elevation of 2700 and 3600 mts. It prefers precipitous rocky slopes covered with dense scrub growth and stunted forest. The thar too lives in herds whose size may be upto 40. In winter they descend to the valley bottom and climb up to higher elevations in search of food in summer.

Goat Antelopes

1. *Seros* (*Capricornis sumatraensis*): A relatively tall animal with a large head, thick neck and short sturdy fore and hind limbs. It is found between an elevation of 2200 and 3300 mts on the slopes of the main Himalaya. They are solitary creatures though 4 to 5 individuals may be seen in the same area. The seros are active both on precipitous rocky slopes and on flatter terrain. This species too is an altitudinal migrant.

2. *Ghoral or Goral* (*Nemorhaedus goral*): This is a stout animal resembling a goat. It has short horns which are not very conspicuous. They inhabit open grassy and rocky areas between an elevation of 2100 and 3000 mts. Their food consists of grasses, leaves, succulent roots and shoots.

Antelopes

Indian antelope or Black buck (*Antelopes cervicarpa*): This is one of the most elegant animals found in India. It has striking colour and beautiful spiralled horns. The coat is of a dark shade and begins to turn black after an age of about 2 or 3 years. In Kumaun, the black buck is found only in the terai and bhabar tract. They avoid dense forests and hilly tracts, preferring to stay in the open grasslands near cultivated fields. There may be between 20 to 30 members in a herd. They move from one place to the other in search of food.

Deer Family

1. *Barasingha or Swamp deer* (*Cervus duvauceli*): This is a stout, beautiful member of the deer family. The colour of its coat varies from brown to yellowish brown, being of a highter shade in summer. Antlers are well developed and form a distinctive pattern. In Kumaun it is restricted to the terai and bhabar tract where it remains in the vicinity of rivers, streams, lakes, swamps and marshes.

The barasinghas begin to feed early in the morning and continue to do so till around mid-day. Thereafter they feed again in the evening hours. They move in small herds whose size varies from area to area. When alarmed the whole herd flees away within a very short period.

2. *Sambhar or Sambar* (*Cervus unicolor*): The sambhar is the largest member of the deer family found in Kumaun. A full grown male may attain a shoulder height of upto 1.6 mts. It has a coarse brown coat with a yellowish or greyish tinge, while the underparts are of a paler shade. Old males are black or almost black. The antlers are

stout and rugged.

In Kumaun, the sambhar is found in the forests and open areas of the foothills. They are also found near cultivated fields that are raised from time to time. The main food of this animal is grass, leaves, succulent shoots and wild fruits. They feed in the early morning and after dusk, preferring to remain hidden in broad daylight. Sambhars are good swimmers and are able to cross swift mountain rivers and streams.

This animal moves and feeds in small herds whose number is not usually more than 10. They move over large distances in search of food.

3. *Hog deer* (*Axis porcinus*): The hog deer is a squat deer resembling a pig and hence the name. Its movements too resemble that of a pig or hog. This animal is small, stoutly built and has a long body and short legs. It has a brown coat which attains a darker shade with age. The underparts and underside of the tail are of a lighter shade.

The hog deer is found in the following tracts of southern Kumaun (in the foothills): (a) grasslands or forests along rivers and streams, (b) open forests along rivers and streams, (c) open grassy plains, and (d) river terraces and river islands.

It is a solitary animal that may move about in twos and threes. The hog deer feeds in the early morning and late evening while the day is spent in the shade of tall grasses and trees. It has a well developed sense of smell, sight and hearing. Their food consists of grasses, leaves, shoots and fruits.

4. *Spotted deer or Cheetal* (*Axis axis*): The cheetal is one of the most beautiful deers found in Kumaun. It has a bright rufous-fawn coat with attractive white spots. The old bucks attain a darker shade with age. The cheetal is found in the terai and bhabar tracts, dun valleys, Siwalik hills and foothills of the lower Himalaya. Their habitat consists of open grasslands along the banks of rivers and streams and in the sal forests. There may be between 40 to 50 animals in each herd which includes 2 or 3 stags. A large pasture may be simultaneously grazed upon by a number of herds.

Cheetal herds feed on grasses, leaves, fruits, succulent shoots, roots and even standing crops in agricultural fields. Feeding starts early in the morning and ends at dusk with a mid-day break in between when the cheetals lie down in the shade of trees to escape from the scorching heat of the sun. They may also raid agricultural fields in the vicinity of forests and grasslands. Cheetal herds move from one area to the other in search of food. Herds of cheetal can be seen in the vicinity of waterholes and open grasslands.

5. *Barking deer or Muntjac* (*Muntiacus muntjak*): This is a relatively small member of the deer family. Its coat is of a dark brown to rufous-fawn colour. The barking deer is found between an elevation of 1500 and 2400 mts. They are solitary in nature and move either singly or in twos and threes. The barking deer prefers to remain in dense forests; though they may come to the periphery of the forest to feed on grasslands.

Their food consists of grass, leaves, wild fruits and succulent shoots. They feed both during the day and after dusk. This animal is an altitudinal migrant and moves to the upper reaches in search of food in summer.

6. *Musk deer* (*Moschus moschiferus*): The musk deer occupies a position between the deer and the antelope. It has a small form, no horns and no facial glands. There is a musk gland in its abdomen. The fresh secretion has an unpleasant odour that attains a pleasant smell when dry. The canine teeth are very well developed in males.

In Kumaun, the musk deer is found above the temperate zone; usually in ones and twos in the sub-alpine and alpine forests. They feed in the early mornings and evenings and lie concealed during the day and night. This species is an altitudinal migrant. They move to areas near the snowline in summer to return to the shelter of the valleys in winter.

The main food of the musk deer consists of grass, leaves, flowers, fruits and lichens.

Wild Pig

1. *Indian wild boar* (*Sus scrofa*): This is a stout pig that resembles the European wild boar. Its coat is relatively thicker and colour is black mixed with grey, rusty-brown or brown. In Kumaun, the wild boar is found in the lower hills upto an elevation of about 1500 mts, inhabiting grasslands, scrub and at times dense forests.

They begin feeding from morning and stop at dusk with a 3 to 4 hour mid-day break in between. Their main food consists of roots, tubers, insects, snakes and crops. The wild boar lives and moves in herds whose size may vary from 5 to 25 or even more.

Pangolin

1. *Indian pangolin* (*Manis crassicaudata*): This is an ant eating

animal found in small number in the terai and bhabar tract. It is nocturnal in habit and spends the day curled up in burrows or amongst rocks and boulders.

BIRDS

Kumaun is the home of thousands of species of birds or avifauna. The bird assemblage of this region has been influenced by the palaearctic, South Indian, Chinese, Tibetan and Central Asian elements. Amongst the larger birds found in Kumaun are kites, eagles, vultures and pheasants. The medium sized birds include fowls, cocks, woodpeckers and swallows while tits and wrablers are some of the smaller birds of Kumaun.

Migratory birds come to Kumaun from the plains of Central India and Central Asia. Altitudinal migration is another form of bird movement. Resident birds move to the higher reaches in summer in search of food and return to the shelter of the valleys in winter.

Different species of birds (both resident and migratory) found in different parts of Kumaun are listed below (after Negi, 1992).

Grebes

1. Great crested grebe
2. Black-necked grebe
3. Little grebe

Pelicans

1. White or rosy pelican
2. Grey or spotted billed pelican

Cormorants and Snake Birds

1. Cormorant
2. Indian shag
3. Little or Pygmy cormorant
4. Darter

Herons

1. Giant heron
2. Grey heron
3. Purple heron
4. Little green heron

5. Pond heron or paddy bird
6. Cattle egret
7. Little egret
8. Little bittern
9. Open-bill stork
10. White-necked stork
11. Black-necked stork

Ibis and Spoonbill

1. Black ibis

Ducks, Geese and Swans

1. Lesser white-fronted goose
2. Whooper swan
3. Lesser whistling teal
4. Spotbill duck
5. Falcated teal
6. Wigeon
7. Garganey
8. Shoveller
9. Common pochard
10. Tufted duck
11. Scaup duck

Hawks and Vultures

1. Black-winged kite
2. Black kite or Pariah
3. Brahminy kite
4. Crested goshawk
5. Besra sparrow hawk
6. Long-legged buzzard
7. Upland buzzard
8. White-eyed buzzard eagle
9. Bonelli's hawk eagle
10. Black eagle
11. Himalayan griffon
12. Indian long-billed vulture
13. Indian white-backed vulture
14. Pied harrier
15. Marsh harrier
16. Short-toed eagle

Falcons

1. Snow partridge
2. Himalayan snow cock
3. Pheasant grouse
4. Chukor partridge
5. Common quail
6. Jungle bush quail
7. Common bill patridge
8. White-crested khalij
9. Red jungle fowl
10. Koklas pheasant
11. Cheer pheasant
12. Common peafowl

Cranes

1. Black-necked crane
2. Demoiselle crane

Rails and Coots

1. Brown crake
2. White-breasted water hen
3. Coot

Sand Grouse

1. Indian sand grouse

Pigeons and Doves

1. Green pigeon
2. Snow pigeon
3. Hill pigeon
4. Wood pigeon or Ring dove
5. Speckled wood pigeon
6. Spotted dove

Parrots

1. Large Indian parakeet
2. Slaty-headed parakeet

Cuckoos

1. Pied-crested cuckoo

2. Indian cuckoo
3. Cuckoo
4. Himalayan cuckoo
5. Indian banded bay cuckoo
6. Koel
7. Large green-billed malkoha
8. Crow pheasant

Owls

1. Barn owl
2. Spotted scops owl
3. Collared scops owl
4. Green horn owl or Eagle owl
5. Brown or Wood owl
6. Short-eared owl.

Nightjars

1. Indian jungle night jar
2. Long-tailed night jar

Swifts

1. Indian edible nest swiftlet
2. House swift
3. Crested swift
4. White rumped spine tail

Kingfishers

1. Lesser pied kingfisher
2. White-breasted kingfisher
3. Blue-throated barbet
4. Crimson-breasted barbet

Woodpeckers

1. Rufous woodpecker
2. Scaly-bellied green woodpecker
3. Large yellow-naped woodpecker
4. Small yellow-naped woodpecker
5. Himalayan golden-backed woodpecker
6. Himalayan pied woodpecker
7. Yellow-fronted pied woodpecker
8. Grey-crowned pied woodpecker.

Larks

1. Bush lark
2. Sand lark
3. Eastern sky lark

Swallows

1. Collard sand martin
2. Swallow
3. Indian cliff swallow
4. Nepal house martin

Shrikes

1. Bay-backed shrike
2. Grey-backed shrike

Orioles

1. Golden oriole
2. Black-headed oriole

Drongos

1. King crow or Black drongo
2. Bronzed drongo

Starlines

1. Black-headed or Brahminy myna
2. Jungle myna

Crows, Magpies and Jays

1. Himalayan jay
2. Indian tree pie
3. Yellow-billed or Alpine chough
4. Jungle crow

Waxwings and Flycatchers

1. Pied flycatcher shrike
2. Small grey cuckoo shrike

Bulbuls

1. Black-headed yellow bulbul
2. Rufous-bellied bulbul

Babblers, Flycatchers. Wrablers and Thrushes

1. Rufous-necked scimitar babbler
2. Scaly-breasted wren babbler
3. Red-billed babbler
4. Striated babbler
5. White-throated laughing thrush
6. Variegated laughing thrush
7. White-spotted laughing thrush
8. Red-billed leiothrix
9. Green shrike babbler
10. Stripe-throated minla
11. Yellow-napped yuhinia
12. Stripe-throated yuhinia
13. Black-capped sibia
14. Rufous-tailed flycatcher
15. Red-breasted flycatcher
16. Little pied flycatcher
17. Small niltava
18. Pale-blue flycatcher
19. Verditer flycatcher
20. Grey-headed flycatcher
21. White-browed fantail flycatcher
22. Paradise flycatcher
23. Dull-salty bellied flycatcher
24. Chestnut-headed ground warbler
25. Strong-footed bush warbler
26. Large bush warbler
27. Aberrant bush warbler
28. Rufous-capped bush warbler
29. Spotted bush warbler
30. Fantail warbler
31. Streaked fantail warbler
32. Franklins' longtail warbler
33. Ashy-longtail warbler
34. Jungle longtail warbler
35. Brown-longtail hill warbler
36. Tailor bird
37. Bristled grass warbler
38. Blyths' reed warbler
39. Brown-leaf warbler

40. Tytlers' leaf warbler
41. Orange-barred leaf warbler
42. Yellow-rumped leaf warbler
43. Grey-faced leaf warbler
44. Large-billed leaf warbler
45. Dull-greenleaf warbler
46. Large-crowned leaf warbler
47. Blyths' leaf warbler
48. Yellow-eyed flycatcher warbler
49. Grey-headed flycatcher warbler
50. Old crest
51. Lesser short wing
52. Blue throat
53. Himalayan ruby throat
54. Orange-flanked bush robin
55. Golden bush robin
56. White-browed bush robin
57. Magpie robin
58. Blue-headed redstart
59. Black redstart
60. Blue-fronted redstart
61. Guldenstadts' redstart
62. White-bellied redstart
63. Black-backed forktail
64. Hodgsons' grandala
65. Little forktail
66. Spotted forktail
67. Purple cocochoa
68. Green cocoluea
69. Hodgsons' bush chat
70. Stone chat
71. White-tailed stone chat
72. Dark-grey bush chat
73. River chat.
74. Indian robin
75. Chestnut-bellied rock thrush
76. Blue whistling thrush
77. Pied ground thrush
78. Orange-headed ground thrush
79. Plain-backed mountain thrush

80. Long-tailed mountain thrush
81. Small-billed mountain
82. White-collared black bird
83. Grey-headed thrush
84. Red-throated thrush
85. Mistle thrush.

Wrens

1. Wren
2. Brown dipper
3. Alpine accentor
4. Altai accentor
5. Robin accentor
6. Rufous-breasted accentor
7. Black-throated accentor

Titmices

1. Grey tit
2. Green-backed tit
3. Crested-black tit
4. Rufous-bellied crested tit
5. Brown-crested tit
6. Yellow-brown tit
7. Red-headed tit

Nuthatches and Wall Creepers

1. Chestnut-bellied nuthatch
2. White-tailed nuthatch
3. Wall creeper
4. Tree creeper
5. Himalayan tree creeper

Pipits and Wagtails

1. Hodgsons' tree pipit
2. Tree pipit
3. Paddy field pipit
4. Red-throated pipit
5. Hodgsons' pipit
6. Brown rock pipit
7. Upland pipit

8. Yellow wagtail
9. Yellow-headed wagtail
10. Grey wagtail
11. White wagtail
12. Large-pied wagtail

Flowerpeckers

1. Thick-billed flowerpecker
2. Yellow-vented flowerpecker
3. Tickells' flowerpecker
4. Plain-coloured flowerpecker
5. Fire-breasted flowerpecker

Sunbirds

1. Purple sunbird
2. Goulds' sunbird
3. Nepal yellow-backed sunbird
4. Black-breasted sunbird
5. Yellow-backed sunbird
6. Fire-tailed sunbird
7. Streaked spider hunter

White Eye

1. White eye

Weaver Birds

1. House sparrow
2. Tree sparrow
3. Cinnamom tree sparrow
4. Yellow-throated sparrow
5. Tibet snow finch
6. Finni's baya
7. Black-throated weaver bird
8. Red munia
9. Common silver bill
10. White-backed munia
11. Spotted munia
12. Black-headed munia

Finches

1. Black and yellow grosbeak
2. Allied grosbeak
3. White-winged grosbeak
4. Spotted winged grosbeak
5. Himalayan green finch
6. Gold-fronted finch
7. Hodgsons' mountain finch
8. Common rose finch
9. Nepal rose finch
10. Pink-browed rose finch
11. Red-mantled rose finch
12. White-browed rose finch
13. Beautiful rose finch
14. Great rose finch
15. Eastern great rose finch
16. Red-breasted rose finch
17. Red cross bill
18. Scarlet finch
19. Gold-headed black finch
20. Red-headed bull finch

Buntings

1. Pine bunting
2. Black-faced bunting
3. White-capped bunting
4. Rock bunting
5. Crested bunting

REPTILES

Kumaun is the home of many species of reptiles. A number of species and varieties of poisonous and non-poisonous snakes are found upto an elevation of about 3000 mts in this region. The gharial and muggar crocodiles live in the Ramganga river in the lower hills. Amongst the smaller four-footed reptiles are the common house gecko, Asian gecko and fan throated lizard.

Snakes

Snakes are found from the terai and bhabar tract to the high hills. Venomous snakes occurring in Kumaun are given below (after Whitlaker, 1975 and Negi, 1992):

1. *Common krait*: Found upto an elevation of about 1700 mts. They are noctural and hide during the day.

2. *Banded krait*: Found in the foothills of Kumaun upto an elevation of about 1500 mts.

3. *Indian spectacled cobra*: Found from the terai and bhabar tract to an elevation of about 4000 mts.

4. *Indian monocled cobra*: Found in the lower hills of Kumaun. They are nocturnal.

5. *King monocled cobra*: Found in the lower hills upto an elevation of about 2000 mts.

6. *Russells' viper*: Found upto an elevation of about 3000 mts. They are capable of moving with great speed.

7. *Himalayan pit viper*: Found upto an elevation of about 4800 mts. It ascends to the highest altitude as compared to any other snake in the world.

The important non-venomous snakes found in Kumaun are:

1. *Himalayan cat snake*: Found upto an elevation of about 3000 mts. They are nocturnal.

2. *Common vine snake*: Found in the foothills and terai and bhabar tracts.

3. *Bronze back tree snake*: Found in the lower hills upto an elevation of about 2000 mts.

4. *Royal snake*: This snake occurs in the foothills, terai and bhabar tracts.

5. *Banded racer*: Found in the terai and bhabar tracts. It hunts for food during the day.

6. *Common rat snake*: This snake is found upto an elevation of about 4000 mts.

7. *Trinket snake*: This snake occurs in the foothills, terai and bhabar tracts upto 2000 mts.

8. *Checkered keelback water snake*: Found along rivers, streams and lakes upto an elevation of about 3000 mts.

9. *Green keelback*: Found in the foothills, terai and bhabar tracts upto an elevation of about 1500 mts.

10. *Striped keelback*: Found in the lower hills, terai and bhabar tracts upto 2000 mts.

11. *Banded kurki*: This snake occurs in the lower hills, terai and bhabar tracts upto an elevation of about 2000 mts. It is nocturnal in habit.

12. *Common wolf snake*: Found in the lower hills upto an elevation of about 2200 mts.

13. *Rock python*: Found in the terai and bhabar tracts and dense forests of the dun valleys.

14. *Common worm snake*: Occurs in the Siwalik hills, terai and bhabar tracts upto an elevation of about 1000 mts. It is a very small snake.

Other snakes found in Kumaun include the following:

1. Slender worm snake
2. Barred wolf snake
3. Mackinnons wolf snake
4. Olive oriental worm snake
5. Cantors' black headed snake
6. Himalayan trinket snake
7. Glossy bellied racer
8. Lesser striped neck snake
9. Himalayan striped neck snake
10. Himalayan sand snake
11. Leiths' sand snake
12. Common cat snake
13. Black cobra
14. Himalayan pit viper
15. Green pit viper

Crocodiles

1. *Gharial* (*Garialis gangeticus*): The gharial is found in and along the rivers of the terai and bhabar tracts. It is found in the Ramganga river.

2. *Common Indian crocodile* (*Crocodilus palustris*): Found in and along the rivers of the terai and bhabar tracts. It is found in the Ramganga river.

Other Reptiles

Other reptiles found in Kumaun include the following:

1. Ganges soft shell tortoise
2. Common box tortoise
3. Starred tortoise
4. Thurgi
5. Common house gecko
6. Asian house gecko

7. Fan throated lizard
8. Common garden lizard
9. Spotted agama
10. Indian slow worm
11. Barred monitor
12. Snake lizard

FISHES

Rivers, streams and lakes of Kumaun abound with fish. Many species of fishes live in the water of this region from the terai and bhabar tract to the high hills. A number of fishes are in the habit of migrating upstream in summer for spawning and breeding.

Important fishes occurring in the waters of Kumaun have been listed in the following text:

1. *Acrossocheilus hexagonolepis*: Found in the rivers and streams of the foothills of Kumaun.

2. *Barilus genus*: This genus consists of the following species:

Barilus barila *Barilus barna*
Barilus bendelisis *Barilus bola*
Barilus vagra

This species are found in the rivers and streams of the foothills, terai and bhabar tracts.

3. *Catla catla*: This fish is found in the water bodies all over Kumaun; in the Nainital and Bhimtal lakes; rivers and streams of the Siwalik, lower and middle Himalaya.

4. *Chagunius chagunio*: Found in the rivers and streams of the terai, bhabar, dun valleys, Siwalik hills and foothills of Kumaun.

5. *Channa gachua*: Restricted to the fresh water bodies of the lower hills. It is not found in the water bodies of higher elevations.

6. *Cirrhinus mirgala*: Restricted to the water bodies of the lower hills, Siwalik hills, dun valleys and the terai and bhabar tracts.

7. *Crossocheilus latius*: Found in the water bodies of all parts of Kumaun, usually upto an elevation of about 3000 mts in the higher hills.

8. *Cyprinus carpo*: Found in the rivers, lakes and streams of the terai and bhabar tracts, dun valleys, Siwalik hills, lower, middle and higher Himalaya.

9. *Danio devario*: Found in the freshwaters of all tracts of

Kumaun; usually upto an elevation of about 3000 mts in the higher Himalaya.

10. *Garra lamta*: Found in the freshwaters of the outer, lower and higher Himalaya of Kumaun. It does not occur at very high elevations.

11. *Glyptothorax telchitta*: Found in various parts of Kumaun; in the rivers Ramganga, Kali and Sarju and in the lakes, e.g. Nainital.

12. *Labeo genus*: The following species of the genus are found in various parts of Kumaun.

Labeo angra
Labeo boga
Labeo calbasu
Labeo goins
Labeo bata
Labeo boggut
Labeo fimbristus
Labeo rohita

13. *Lepidocepalichthys guntea*
14. *Mastacembelus armatus*
15. *Neomacheilus botia*
16. *Puntius amphibius*
17. *Puntius carnaticus*
18. *Puntius conchonius*
19. *Puntius chola*
20. *Puntius sophore*
21. *Salmogairdnerii gairdnerii*
22. *Salmo trutta fario*
23. *Tor tor*.

NATIONAL PARKS AND SANCTUARIES

Kumaun represents the transition zone between the fauna of the western and central Himalaya. It is rich in both flora and fauna. In fact, India's first national park is situated in the lower hills of Kumaun. The national parks and sanctuaries of Kumaun have been described in the following text:

Corbett National Park

This is the first national park of India. It is situated in the lower hills falling within Pauri and Nainital districts. Set up as the Hailey National Park in 1936 after the then Governor of United Province, its name was changed to Ramganga National Park before being christened after Jim Corbett, the famous naturalist in 1956.

At present, this national park encompasses an area of about 520 sq. kms which includes the vast reservoir formed by the Kalagarh dam across the river Ramganga. It is one of the best manged protected areas in the country and attracts thousands of visitors each year. This area is also a tiger reserve under the Project Tiger.

For the purpose of management, this national park has been demarcated into the following: (a) core zone, (b) tourist zone, and (c) buffer zone.

Geomorphology: This park encompasses the Patli dun valley formed by the Ramganga river. In the north lie the foothills of the lower Himalaya while the low rolling Siwalik hills are in the south. The Ramganga river flows in between these two hill ranges. A number of seasonal streams locally known as *sots* drain this tract.

Climate: A typical sub-tropical climate prevails in the Corbett National Park. The SW monsoons cause heavy and widespread rains from July to September. Autumn is sunny and fair. Winters are cold and frost commonly occurs during this season. It becomes very hot in this area during the summer months and the maximum temperature may be about 38°C.

Flora: The principal forest types found in this national park are listed below:

1. Moist and dry sal forest
2. Khair and sisham riverine forest
3. Chir pine forest
4. Moist deciduous scrub and grassland.

Fauna: This national park is the home of the tiger whose numbers have gone up due to strict protection measures. Many elephant herds also live in this park while others migrate to and from the adjoining tracts. Other mammals found here include the leopard, jackal, wild boar, sambhar, spotted deer, hog deer, barking deer, Hanuman monkey and langur.

Amongst the birds found in this park are the jungle fowl, wrablers, tits, pheasants and woodpeckers. The vast reservoir is the home of many resident and aquatic birds. The reptiles include the gharial, muggar, Indian rock python, king cobra, cobra, Russells' viper and common krait.

2. Ascot Musk Deer Sanctuary

This is a 284 sq. km sanctuary located in Pithoragarh district of the higher Himalaya of Kumaun. This sanctuary has been set up primarily with the object of conserving the musk deer and its habitat.

Geomorphology: The upper tracts have been shaped by the action of glaciers while running water has carved the terrain of the lower reaches. The prominent geomorphic features of this sanctuary include glacial amphitheatres, ridges, spurs, V-shaped valleys and river cut and river built terraces.

Climate: The climatic conditions vary from temperate to arctic. Summers are mild in this tract. However, winters are very severe and a thick layer of snow covers the ground for several weeks at a stretch during the cold season.

Flora: The main forest types found in this sanctuary are listed below:

1. Ban and Moru oak forests
2. Moist deodar forest
3. Temperate mixed coniferous forest
4. Temperate mixed deciduous forest
5. Kharsu oak forest
6. Fir and spruce forest
7. Blue pine forest
8. Dry temperate mixed forest
9. Dry deodar forest
10. Sub-alpine birch fir forest
11. Birch-rhododendron scrub forest
12. Moist alpine scrub forest
13. Dry alpine scrub forest

Fauna: The Askot sanctuary is the home of the must deer. Intensive efforts have been initiated to conserve this elusive species. Other mammals found in this sanctuary include the leopard, jungle cat, civet cat, barking deer, serow, ghoral and brown bear. Many species of high altitude birds are also found in this sanctuary.

3. Binsar Wildlife Sanctuary

The Binsar wild life sanctuary encompasses an area of about 50 sq. kms in the middle Himalaya near Almora. This tract is considered to be one of the most beautiful tracts in Kumaun.

Geomorphology: The slopes vary from steep to very steep. The terrain has been shaped by the action of running water.

Climate: The climatic conditions prevailing in the Binsar sanctuary range from temperate to sub-arctic. Winters are very cold and heavy snowfall is received.

Flora: The main forest types found in this sanctuary are listed below:

1. Oak and moist mixed deciduous forests.
2. Temperate moist coniferous forests including deodar and blue pine.
3. Sub-alpine birch, fir and spruce forests.
4. Sub-alpine pasture and scrub.

Fauna: This sanctuary is the home of many high altitude species of animals and birds which include the leopard, civet cat, serow, gharial, musk deer, brown bear and khalij pheasant.

8

History

Kumaun has an interesting and long history which can be traced from the earliest times to the present. Its history has many common features with that of the adjoining region of Garhwal, particularly in the early ages. The history of Kumaun has been discussed in brief in this chapter.

Prehistory and Protohistory

Human beings left their imprints in Kumaun in the perhistoric times. This is borne by the following features:

1. Stone age tools have been found at different sites in Nainital and Almora districts.

2. Painted rock shelters are found in the Binsar Gad valley, viz. the famous Lakhu Udyar rock shelter. There are believed to be of the mesolithic-chalcolithic period.

3. There are depressions of varying sizes and shapes, found all over Kumaun. The smaller ones are ritualistic and sepulchral. On the other hand the larger depressions have a depth and diameter varying from 1 to 2 mts. They may be dwelling pits.

4. Cists vary in size and shape and are found on terraces in the valleys. They are made up of four orthostats laid under the ground and covered by a single stone slab. In some localities they may be raised in the row of 3 to 5 or even more. In such cases there are common inner orthostats which serve the purpose of partition.

Pots of different shapes and apparent uses are found inside these cists. The pottery is either red or grey and is wheel-turned.

5. Depressions and cists are found in many parts of Kumaun, e.g. Katyur, Takula, Someshwar, Kaidarau and Kakadighat. This indicates

that human beings lived in all these parts during the pre and proto-historic times. The makers of depressions were probably migratory hunter-gatherers who lived in groups and those of cists were food-growers.

Early History

Kuninda Period

This period synchronises with the rise of Kunindas whose coins are found and about whom there is mention in our scriptures like the Mahabharat, Ramayan and Puranas. The Kunindas reigned supreme in Kumaun from the time of Panini (500 BC) to that of Varahmihira (6th century).

Coins: Silver and copper coins were issued during this period. These may be divided into three series:

a) Amoghabhuti type issued in silver and copper and bearing the name of King Amoghabhuti in Brahmi and Kharoshthi legends.

b) Almora type issued in silver and copper and bearing the names of a number of Kings in Brahmi legend.

c) Chatreshvar of anonymous type issued in copper and bearing the name of Chatreshvar (Lord Shiva).

Society: During this period, the societal set up of the people living in Kumaun was tribal in character. The people lived in communities that were widely distributed. They used wheel-turned pottery in various forms and sizes. Their houses were of brick, which were raised over a foundation made of river bed pebbles. Iron was forged and used for making fish-hooks, nails and arrow heads. Combs and bangles were made of bones.

Agriculture was the mainstay of the people in this period. They also reared sheep and goats whose hairs was woven into various forms of cloth. These people were also in the habit of eating meat and fish.

Religion: The people of the Kuninda period believed in Lord Shiva whom they worshipped in symbolic form as Chhatra and later in the anthropomorphic form. At a later period their state was dedicated to Lord Shiva. The Kunindas also performed vedic sacrifices as is evident from the brick inscriptions.

Trade and Commerce: Trade and commercial activities had attained considerable development during this period. These people were in close contact with societies outside Kumaun. The inhabitants of the border areas traded with Tibet during and after the Mauryan times. Silver is believed to have been obtained in barter from Tibet. This is evidenced from the fact that many silver coins were in vogue during the Kuninda period.

After the emergence of Samudra Gupta, the Kunindas slowly declined. As their dominion broke up, one of the branches set itself up at Katyur and this saw the advent of the Katyuri period. Besides the main branch, many small chiefdoms also came up in these hills.

Katyuri Period

The early medieval history of Kumaun centres around the Katyuris. This name is derived from the Katyur valley of Almora district which in turn may be the Kartripur of Prayaag Prashasti. The Katyuris ruled from the seventh to the eleventh centuries. At the peak of their powers, they ruled over whole of Kumaun, Garhwal, parts of western Nepal and eastern H.P. For most of this period their capital was at Kartikeyapur (modern Baijnath, Almora).

Culture and Art: During the early part of this period, bricks replaced stone. Even very large structures like temples and forts were made from stone whose huge blocks were apparently transported over very large distances. It is clear that the people of this period were well versed in the art of stone quarrying and transport. Iron tools were locally made and used even for very heavy work. Clamps of iron were used to bind the slabs of stone together.

The salient features of the culture and art during this period are given below:

a) Dance, music, gladiatorial combats, hunting and drinking formed the main source of recreation and sport.

b) Pottery, craft was very well developed during the Katyuri period.

c) The famous Katyuri art which found its expression in many temples and sculptures developed during this period.

d) Many festivals and fairs were organised by the people during this period.

Economy: The economy was largely based on agriculture. Some people were engaged in goat and sheep rearing and others in trade and commerce with the people of the plains and with Tibetans. Other vocations included weavers, dyers, tanners, cobblers, tailors, masons, sculpters, carpenters, iron smiths, priests and soldiers. In many areas these people lived within their own community or villages dominated by them.

Religion: Lord Shiva was worshipped very widely in Kumaun during this period. Other gods and goddesses revered by the people included Shakti, Vishnu, Surya, Ganesh, Brahma, Kuber and Ganga. The impact of other religions or faiths was minimal in the Katyuri times.

Administration: During this period there was an elaborated system of administration in Kumaun. The administrative set up drew inspiration from the Gupta empire.

Medieval Period

The Katyuri kingdom declined in the eleventh century and there emerged a number of smaller principalities which severed relations with the central powers. This created a political vacuum and the Mallas of Nepal subjugated parts of Kumaun during the twelfth and first half of the thirteenth centuries. However, the Mallas could not retain territory in Kumaun for long and there emerged a number of powerful principalities in different parts of Kumaun. These have been listed below (after Joshi, 1988).

1. Later Katyuris of Pali region (Ranikhet, Almora)
2. Later Katyuris of Katyur valley (Almora)
3. Raikas of Doti-Dandeldhura (western Nepal)
4. Raikas of Sira (Didihat, Pithoragarh)
5. Palas of Askot (Askot, Pithoragarh)
6. Mankotis of Gangolihat (Pithoragarh)
7. Bams of Sor valley (Pithoragarh)
8. Chands of Kali Kumaun (Champawat, Pithoragarh)

The Chands and Raikas were relatively more powerful and ambitious. They dominated the other kingdoms and gradually the Chands emerged as the most powerful principality in the whole of Kumaun during this period. They virtually brought the entire region under their control.

The medieval history of Kumaun synchronises with that of the Chand dynasty. This history has been traced in the following text.

Chronology: The origin of the Chand dynasty is still a matter of debate amongst historians. According to one school of thought, Som Chand from Jhansi or Kannauj was the founder of this dynasty. Another theory suggests that Thohar Chand was the founder. He came to Kumaun in 1261. His line became extinct after the death of his great grandson after which Gyan Chand, a direct descendant of Thohar Chand's uncle was crowned at Champawat. The Chand dynasty prospered during and after the rule of Gyan Chand. The Gorkhas invaded Kumaun in 1790 during the rule of Mahendra Chand Singh after which this dynasty was terminated. The capital of their state was shifted to Almora by Bhishma Chand. The rulers of the Chand dynasty were:

Abhay Chand; Gyan Chand; Vikram Chand; Bharati Chand; Rattan Chand; Kirti Chand; Pratap Chand; Bhishma Chand; Kalayan Chand-I; Ajit Chand; Kalyan Chand-II and finally the last ruler Mahendra Chand Singh.

Administration: During the medieval period, Kumaun was made up of chiefdoms which were controlled by one central authority. These were headed by the chief who was assisted in administrative matters by distinguished men from different walks of life. Names were assigned to various professional groups, which later became accepted surnames.

The chiefs and his advisors enjoyed a considerable degree of autonomy. These people owned vast tracts of land and were frequently bestowed with honours and awards. The Chands did not maintain a regular standing army and depended on the chiefs to provide them with warriors in times of war and emergency.

Economy and Vocations: The salient features of the economy and vocations of the people of Kumaun during this period were:

a) Agriculture was the mainstay of the economy. Even the chiefs had agricultural land for their cultivation.

b) Cattle rearing, sheep farming and production of woollen garments were also practised.

c) Carpentry was well developed during this period. The old buildings at Bageshwar, Champawat, Dwarahat and Almora testify to this.

d) Merchants were known as *sahus* who later came to be known as the Sah's.

e) People living in the northern areas traded with Tibet. They travelled in caravans to places upto Lhasa.

Religion: Lord Shiva was the main God worshipped during this period. Other dieties included Vishnu, Shakti, Brahma, Naga, Ganesh and Yaksha. Different forms of Bhairav were also worshipped by the people. In the latter part of this period, worship of Rama, Krishna and Hanuman became more prevalent.

Kumauni Language: The Kumauni language was fully developed during the medieval period. All the ruling dynasties of this period, viz. the later Katyuris, Palas, Raikas, Bams, Mankotias and Chands used this as their official language. Royal charters were regularly issued in Kumauni.

Gorkha Rule (1790 to 1815)

The last ruler of the Chand dynasty Mahendra Chand Singh was very weak. The Gorkhas were the natives of western Nepal. The Gorkha ruler Rana Bhadur invaded and captured Kumaun with the help of Harsha Dev Joshi who was the Prime Minister of the ruler at Almora. He appointed Joga Malla Sah as the Governor of Kumaun. They were later succeeded by Kazi Nar Shahi, Ajab Singh Thapa, Bam Shah, Dhaukal Sur Sing, Kazi Gazeshwar and Rituraja Thapa.

Administration: The Gorkha rule of Kumaun was mainly in the hands of the military officers. They ruled Kumaun with a heavy hand and punished the local people arbitrarily. They had a whimsical attitude towards their subjects.

The dispensation of justice was done in the following ways:

a) Many cases were decided on the basis of oral examination by military officials.

b) Priests decided land dispute cases by picking pieces of paper.

c) Often poison was administered to an accused and if he survived, he was taken to be innocent.

d) Murder was usually punished by hanging to death from trees. Brahmin murders were banished from the state.

e) Cow slaughter was punishable with death.

Revenue system: The system of revenue and taxes was fully altered during the Gorkha rule. They introduced several new taxes. Slave trade was also practiced by the Gorkha rulers and more than thirty thousand slaves were sold in a year to earn more revenue. The property of those who could not pay taxes was taken over by the regime.

Collection of heavy fines was another source of revenue during the Gorkhá rule in Kumaun. As a result of these harsh taxes and repressive rule, the condition of the economy deteriorated steadily leading to misery and poverty.

British Rule (1815 to 1947)

Annexation of Kumaun

Kumaun became a part of British India in 1815 after the treaty of Sagauli. Two factors caused the annexation of Kumaun by the British:

1. After capturing Kumaun, the Gorkhas raided the plains and often came in conflict with the British.

2. On the other hand the British too were attracted towards the hills and wanted to bring Kumaun under their rule. The climate of this mountainous region was more suited to them. They wanted to establish garrison towns where their troops could rest in times of peace.

As a result, the interests of the Gorkhas and the British clashed and when negotiations failed, Lord Hastings, the Governor-General of India ordered his forces to attack the Gorkhas in 1814. Maj. Gen. Gillepse was sent with a strong force which attacked the capital Almora from the west. It was captured on 27th April, 1815 after two days of bitter fighting.

The treaty of Sagauli was signed and Kumaun was ceded to the British. Gardner was made incharge of this mountainous region. He laid the foundation of administration in Kumaun.

Administrative Reorganization

The following administrative reorganization was carried out by the British in Kumaun.

1. In 1839, Kumaun was divided into two districts, viz. Kumaun and Garhwal. A district named Terai was created in 1842.

2. As a result of further reorganization in 1892, new districts of Almora, Nainital and British-Garhwal were created.

3. In 1857, the headquarters of Kumaun division were moved to Nainital.

Agriculture

Agriculture had deteriorated to very low levels during the Gorkha rule. However, things began to improve in the British regime. The total annual production went up by about 15% from 1815 to 1846. Later on cotton cropping expanded very rapidly with the opening of the rail line. A network of canals was constructed in the terai areas for irrigating agricultural land.

In 1820 itself, the settlement record of rights of virtually every village was formed and a detailed system of land records formulated.

Trade and Commerce

People living in the northern areas (e.g. Bhotiyas) engaged in trade with Tibet. They exported foodgrains, soaps, oil, tobacco, textiles, tea, sugar and in exchange obtained salt, borax, raw wool, Tibetan tea and gold. Thousands of people on both sides of the border were engaged in this lucrative trade. In fact, this was later used by the British to further their interests in Tibet.

In the interior areas of Kumaun, trade and commerce were mainly carried out at fairs at convenient centres like Jauljibi, Thal and Bageshwar. Many commodities were exchanged with the neighbouring areas but the importance of these fairs has slowly declined.

Forest Exploitation

In the early years of the British rule in Kumaun, emphasis was on heavy exploitation of the forests with scant regard for their conservation and regeneration. Sal forests were extensively felled in the terai and bhabar areas. However, scientific management of these forests was introduced towards the end of the last century. Forest boundaries were clearly demarcated and the clearing of forests for agriculture was more or less stopped.

In 1917, the forests were divided into three categories, viz. the Reserved Forests managed by the Forest Department; the Protected or Civil Forests under the district administration and the Panchayati Forests under the management and ownership of the village societies. However, this system did not work well and the people launched an agitation under the Kumaun Association to deal with their forest related problems.

Road and Rail Communications

The rail line was extended upto Kathgodam in 1884. Earlier in 1872, the first cart road was commissioned linking Ramnagar with Ranikhet. In 1907, a rail line from Moradabad was extended upto

Ramnagar. During this period Kathgodam was connected by road with both Nainital and Ranikhet (from where it was extended to Almora) Soon afterwards many interior areas were linked by road and this led to the development of the Kumaun hills.

Industrial Development

Kumaun has been an agricultural land ever since human civilization appeared here. With the advent of the British on the scene, industries began to develop in Kumaun too.

1. *Tea Industry*: The first tea nurseries were developed near Bhimtal and Almora in 1841. In the next four decades tea gardens had come up at Bhowali, Bhimtal, Gorkhakhal, Ramnagar, Almora and Berinag. The tea from Berinag became very popular. However, it was replaced by the less costly tea from Assam in the early part of this century.

2. *Iron Industry*: In 1860 a private British firm began to manufacture iron from local raw material near Khurpatal. The government allotted forests to this company as coal was not readily avilable. However, this company could not last for long due to the development of iron plants in Bihar and West Bengal.

3. *Small Scale Industries*: Many small scale industries developed in different parts of Kumaun during the British period. This included blankets, woollen garments, carpets, baskets and handicraft industries.

Freedom Struggle

1. *Mutiny of 1857*: The people of Kumaun remained largely aloof during the national uprising of 1857. In fact many of them even showed their loyalty towards the British. This was due to the popular governership of Henry Ramsay.

2. *Early Movements*: However, the cordial relations between the British masters and the people of Kumaun did not last for long. There began movements for redressal of grievances in the early part of this century, much under the influence of nationalistic movements in the plains.

The *Almora Akhbar* that began publication in 1871 became very strong in its criticism of the British. It was banned as a result of its anti-establishment stance in 1918. A newspaper named *Shakti* started publication in the same year. These periodicals helped to kindle nationalistic feelings in the people of Kumaun. They also became a vehicle for pro-independence poets and authors like Gumani, Mola

Ram, Gaurda, H.C. Joshi and B.L. Sah.

3. *Anti-Kuli Begaar Movement*: Kuli-Begaar was a system under which the local people were supposed to serve as porters to British officials on tour to different parts of Kumaun and also to supply free ration to them. This was a source of humiliation to the local people.

The Kumaun Parishad which was formed in 1916 to create social, political, economical and cultural awareness amongst the Kumaunis, took up this issue at its Haldwani session in 1920 under the presidentship of T.D. Gairola. It condemned this system and demanded that the U.P. government abolish it with immediate effect.

Protests were held in different parts of Kumaun, mainly at Almora and Haldwani. A large meet was organised at Bageshwar during January 1921 in which stalwarts like B.D. Pande, H.G. Pant and Chiranji Lal participated. All registers carrying the names of the labourers were forcibly thrown into the Sarju river. Thus, victory was achieved and this brought the Kumaunis in the mainstream of the freedom struggle of India. Mahatma Gandhi described this unique agitation as a bloodless revolution.

4. *Elections to State Assembly*: Kumaun was a part of the state of United Provinces. In 1923, the first elections to the legislative assembly of U.P. were held. Govind Ballabh Pant was elected from Nainital and Har Govind Pant from Almora. The former became the leader of the Swaraj Party.

They actively participated in the proceedings of the assembly. Due to the efforts of G.B. Pant, the non-regulation system prevalent in Kumaun was withdrawn in 1925. Later on, in 1928, the Naik reform act was passed vide which a ban was imposed on the prostitution of teenage girls of the naik community of Kumaun and they were provided with educational facilities.

5. *1930 to 1942*: During the famous Dandi march of 1930, many people of Kumaun also joined Mahatma Gandhi. Many processions were taken out by people in towns like Almora, Nainital and Haldwani between 1930 and 1934 during the civil disobedience movement. Hundreds of them were injured or arrested as a result.

When elections were held in 1937, G.B. Pant and H.G. Pant were elected from Nainital and Almora respectively. The former was elected leader of the Congress Party and became the chief minister of the province.

6. *Quit India Movement*: The Quit India Movement which started in 1942 also shook the Kumaun hills. The people took active part in

this nation wide movement which finally led to the collapse of the British rule in India. As the top leaders of Kumaun too had been jailed in the very beginning, the younger leaders took over. In some places there was violence as the protesters cut telegraph and telephone lines; attacked police stations, set fire to post offices and resin depots. Women joined men in this movement on a large scale. The people protested very vehemently in the rural areas and police had to open fire in places like Deoghat and Salt. This helped to throw off the British yoke from India.

India gained independence in 1947. Since then Kumaun is an administrative division within the state of Uttar Pradesh. It has made rapid strides in development and self-reliance.

9

People and Culture

PEOPLE

Like most societies of the world, the initial society of Kumaun too was of hunter-gatherers which in due course of time changed to nomad-pastoral, pastoral-agrarian, pastoral trader and finally to agrarian-pastoral or agrarian. These stages are clear as people first settled in the valleys, then moved to the mountains and then there was a reverse flow in which people finally settled in the malaria prone terai and bhabar tract.

Evidence available in the Siwalik hills of Kumaun indicates that the first group of settlers came to these hills from the southern foothills. The next arrivals may have been from the east (Nepal); west (Garhwal, Himachal) and finally from Tibet in the north. In fact, Kumaun is a melting pot of many ethnic groups which came from different parts of Asia and settled in this hilly region.

Ethnic and Social Layers

The ethnic and social layers of the people of Kumaun is given in the following lines.

1. *Aryans*: Different waves of Aryans settled in Kumaun. They speak the Indo-Aryan language and are divided into sub-groups or castes such as Brahmins, Rajputs and Khasas. These people are of central or west Asian origin though they may not have come to Kumaun directly from these areas.

2. *Mongoloids or Tibeto-mongoloids*: These people are primarily mongoloids and inhabit the northern frontier lands of Kumaun. They are made up of different sub-groups such as the jadhs, tolchas, marchas, jauharis, darmis, nyansis, chaudansis, baurajis, tharus,

boksas and mihars. This ethnic group speaks Tibeto-Burmese or naag language. They came to Kumaun from parts of the eastern or southern China or eastern Tibet.

3. *Negroids*: These people resemble the inhabitants of Africa. They originated in south Africa, Arabia or south-west India though they may not have come to Kumaun directly from there. The negroids are made up of the Kols or Shilpkars. They speak Kol or Munda or a Austro-Asiatic language.

Ethnic Groups

The people of Kumaun are made up of a number of ethnic groups. These are:

1. *Kols or Munda group*: They are the first known ethnic group of Kumaun. This group is an off shoot of the negroid group who now reside in the coastal tracts of south India, the Andaman islands and Australia. It is believed that during the paleolithic and neolithic ages, these people were defeated by the Dravidians and forced to move to parts of the Deccan plateau or the Himalaya.

Later on other ethnic groups settled in Kumaun. The imprint of the Kols or Mundas is found on all these groups which assimilated the original settlers of this hilly region.

The present day representatives of the Kols are the shilpkars who founded the initial society of Kumaun. They spoke a Munda dialect and many words of this dialect are still used in most parts of the western Himalaya. As a matter of fact the names of many rivers, streams, mountains and villages of Kumaun show the influence of the ancient Munda dialect. Today the Kols or Mundas occupy the lowest rung of the hierarchy in Kumauni society.

2. *Kiraats or Mongoloid group*: This is considered to be the second oldest ethnic group of Kumaun. They are the nomad-pastoral Mongoloid Kiraats who originated from eastern or southern China or eastern Tibet. The different sub-groups are the bhil-kiraats, naags, tangans, palangans and kinnars who came and settled in Kumaun at different times or periods. They also find mention in the vedas. This ethnic group mainly inhabits the northern, north-western and north-eastern parts of Kumaun.

The present day representative sub-groups of the Kiraats inhabiting various parts of Kumaun are:

a) Jauharis, darmis, nyansis and chaudansis of the border areas of Pithoragarh.

b) The baurajis of Askot.

c) The rauts of Lohaghat.

d) The tharus and boksas of the Terai.

The present condition of the Kiraats has been influenced by the environment in which they live. Those living in the frontier areas adopted characters like nomadism and the facial features of the Tibetan communities. Groups living in relatively inhospitable areas have been able to preserve their original ethnological characters to a considerable extent.

3. *Khasa groups*: This is the third and perhaps the most important ethnic group of Kumaun. It consists of three sub-groups, viz. Armenoids, Alpanoids and Dinaris. They are the descendants of the early Aryan settlers, but probably are not related to the vedic Aryans.

At present about 80 to 90% of the population of Kumaun and the adjoining regions is of Khasa origin. Even the Buddists, Muslims or Christians of this region were Khasas in the distant past. Later on they converted into these religions. In the beginning the Khasas were nomad-pastoralists. They became agriculturists after a few centuries. The present day caste system is also a relatively recent development which did not exist in the early stages.

It is believed that the Shakas and Yavans came to Kumaun from Shakistan in Iran during the second and third centuries B.C. The graves found at Malari, Baijnath and Dwarahat are believed to be of these settlers. The process of arrival may have taken more than 3 to 4 centuries.

The Khasas took to nomad-pastoralism and later on settled down to agriculture. They were also good fighters and joined the armies of the rulers. After the British annexation of Kumaun, they joined the British army. At present they are ably serving in various capacities in the Indian army.

This ethnic group initially worshipped the sun god in various forms. They built sun temples in many parts of Kumaun. Later on they took to worshipping Shiv, Parvati, Laxmi, Durga, Vishnu. Thousands of temples were built by them in many parts of Kumaun. A large number of such temples are surviving even today.

4. *Aryans*: The Vedic Aryans are believed to be the latest ethnic groups of mass settlers in Kumaun. A stream of Buddhists of Tibet went through Kumaun. The influence of Aryan settlers is seen in many parts of this region.

During the rule of Bindusar, the Khasas rose in revolt and Ashoka had to come to pacify them. The Chinese traveller Yuan Chwang who visited this region in the seventh century has described the advent of the Aryans in Kumaun.

Caste Stratification

In the beginning there was virtually no caste stratification in the Kumauni society. Settlers from the plains brought distinct castes with them. The main castes of Kumaun are:

1. *Brahmins*: The Brahmins are the priests and their descendants. They are believed to be close to God. The Brahmins in turn are made up of the following sub-divisions, viz. Chauthani, Sarola, Khasi, Pitali or Nanagotri, Beda.

2. *Rajputs*: The Rajputs are the warrior caste. They in turn are made up of the following sub-divisions, viz. Asal Rajputs or Kshatriyas, Khas Rajputs or Khasi jamidars. The first category consists of the later immigrants who settled in Kumaun after running away from the Muslim rulers in the plains. The second category emerged from the original inhabitants of Khasa and Kiraat groups.

3. *Shilpkars or Doms*: They are the lowest rung of the Kumauni society. There are two main sub-divisions of this caste, viz.

a) *Khalaits*: Orhs, Kolis, Chamars, Tamtas, Lwars, Dhalautis, Dholis, Dhonis, Bakharis, Aujis, Bajjis and Bhuls.

b) *Mangkhanis*: Badis, Bedas, Bhats, Mirasis and Hurkias.

The proportional distribution of the religious groups in Kumaun is given in the following table:

Class/Caste	*% of total*
Immigrant Brahmin	03
Khasi Brahmin	20
Thakur Rajput	13
Khasi Rajput	29
Shilpkars or Doms	25
Others	10

Slave Trade

Conflicts and compromises between different ethnic groups in the society in Kumaun gave rise to the oppressive system of slavery which continued till as late as 1815 before the advent of the British on the scene. The salient features of this unwanted and unfortunate practice were:

1. In the past the privileged class mantained two types of slaves, viz. one for household work and the other for the cultivation of their land. In many areas the so called ownership of slaves passed on from father to son and this was a heriditary ownership.

2. During the British period, the lower classes were forced to do compulsory and unpaid work (*begar*) for the officers of the crown and in some cases for upper class Indians.

3. When multi-state feudalism was at its peak in Kumaun, the level of taxes became so high that the poor people were forced to sell their children into slavery.

4. Money-lenders too contributed to this system. When a man could not repay a loan he was forced into slavery. In case the original borrower died, the social system forced slavery upon his son.

5. This baneful system of slavery reached its peak during the Gorkha rule in Kumaun. It has been estimated that as many as 1 lakh slaves were sold every year. This included men, women and even children. They were forcibly taken to Nepal. In the interior areas a human being fetched a price lower than that of a horse or camel. The price of human being ranged from Rs. 15 to Rs. 150 depending upon the condition of the slave and the area where the sale was taking place.

Some places became centres for slave trade, e.g. Kashipur, Rudrapur and Haldwani.

KUMAUNI LANGUAGE AND DIALECTS

Kumauni is the chief language spoken in Kumaun. It is one of the constituent languages which form the central pahari group. The word pahari language in turn denotes all the Indo-Aryan languages spoken in the southern slopes of the western Himalaya from Jammu to Nepal. These languages may be sub-divided into the following:

1. *Western Pahari*: Spoken in parts of the Jammu hills and H.P.
2. *Central Pahari*: Spoken in the eastern parts of H.P. and Garhwal.
3. *Eastern Pahari*: Spoken in the eastern parts of Garhwal and in Kumaun.

In the past the region around Champawat in Pithoragarh was known as Kumu and the language spoken there was called Kumaiyya. As the domain of the rulers of this region expanded, the entire territory came to be known as Kumaun and the main language and its dialects were named Kumauni.

Influence of Other Languages and Dialects

Many languages and dialects have had an effect on this language. These are:

1. Bhotiya dialects of the Tibeto-Burman family in the north (Tibet and China)
2. Hindi or Hindustani or Sanskrit in the south
3. Nepali in the east
4. Garhwali in the west.

People speaking these languages and dialects moved across the borders of various regions and hence the languages were influenced by each other.

Origin of Kumauni

In the historical past Kumaun was inhabited by people of Aryan and non-Aryan origin. They brought various languages and dialects with them and this led to the origin of the Kumauni language. Different authorities have advanced many theories to explain the origin of Kumauni.

The most widely accepted theory is that the origin of Kumauni is Sauraseni Prakrit which is also the source language for Rajasthani and Brij Bhasha. Another theory advocates that the pahari language spoken in other parts of the western Himalaya is the source of Kumauni.

Dardo-Pahari origin: During the Mauryan period there is no evidence of the spread of the Aryan tongue in this region. It may be possible that the Dardic speaking Aryans were still in the process of settling in other parts of the western Himalaya in the Mauryan times. When the Aryans came to Kumaun they superimposed their tongue on that being used by the original settlers of this region.

The base of the Kumauni language was perhaps an Aryan tongue of the Khasas which is known as the Udican. It was probably Sanskrit based. Experts are of the opinion that many present day languages and dialects of the western Himalaya are Udican based, viz. Kashmiri, Ladhaki of Ladhak, Khowar of Chirtal and Shina of Gilgit. The languages spoken in the southern slopes of the western Himalaya bear resemblance to the language of the Khasas.

As a matter of fact the dardo language is also the base of

Garhwali, Nepali, Bengali, Assamese, Oriya and Rajasthani. These languages are in many ways different from those of the plains of northern and central India.

Pahari languages: Kumauni bears close resemblance with central and eastern pahari languages and dialects, e.g.

a) Sirmauri of Sirmur
b) Mahasui of upper Shimla
c) Kului of the Kullu valley
d) Garhwali of Garhwal
e) Mandiali of Mandi and Sundernagar
f) Bilaspuri of Bilaspur
g) Kangri of the Kangra valley
h) Chambiali of Chamba.

Morphological correlations: All pahari languages have a number of distint correlations, e.g.

1. The ending—*i* is used as a conjunctive participle to express the completive aspect of a peripharatic construction.

2. These languages use—*la* either for all persons or in the second and third persons.

3. The future marker—*lo* does not usually undergo change for number, gender and person.

4. Most pahari languages attest a close affinity in causative formative.

5. The continuative aspect is expressed with the help of a number of auxiliaries.

Development of Kumauni

Most experts have identified the following three phonological and morphological stages in the development of Kumauni:

1. *Proto-Kumauni*: This is the same as Dardo-Pahari. It has a common vocabulary with other groups.

2. *Old-Kumauni*: The main features of old-Kumauni are listed in the text given below:

a) Preservation of word final vowels.
b) Maintenance of both quantity and quality of the penultimate vowels.
c) Non-gladalization of *e* and *o* occurring in a stressed syllable (due to a musical accent).
d) Absolute preservation of vowel sequences and dipthongs.

3. *Modern-Kumauni*: The main features of modern Kumauni are listed below:

a) Most of the vocabulary of this language originated from old and middle-Indo-Aryan languages spoken by the Khasas.
b) The proportion of non-Aryan vocabulary is small.
c) A number of words used during the first century of the Christian era are in vogue in modern Kumauni.
d) The linguistic elements bear close resemblance with the Dardo-Pahari languages.
e) Many words belonging to the Tibeto-Burman dialects are now used in modern-Kumauni.
f) There are occasional words of Dravidian languages in modern Kumauni.

Dialects of Kumauni

There are many dialects in Kumauni. These differ from area to area. The chief dialects of this language are listed below:

1. Khasparjiya
2. Danpuriya
3. Phaldakoti
4. Pachhai
5. Gangoli
6. Kumaiya
7. Bhabari
8. Darmiya
9. Johari
10. Chhakhatiya
11. Askoti
12. Sirali
13. Soryali
14. Rau-chaubhainsi
15. Chaugarkhiya

Kumauni language has a vast wealth of literature in the form of prose, poetry and folk songs.

RELIGION

A majority of Kumaunis are Hindus. Others include Muslims, Sikhs, Jains, Buddhists and Christians. The religion-wise proportional distribution of the population in Kumaun is given in the following table:

Religion	*% of total population*
Hindus	91.5
Sikhs	4.0
Buddhists	2.5

Muslims	1.0
Others	1.0

Gods: Lord Shiva is widely worshipped in Kumaun hills. The road to the holy abode of Lord Shiva at Mansarovar in Tibet runs through Kumaun. His wife Shakti, Parvati or Uma is believed to be the daughter of the Himalaya. There are thousands of temples of all sizes devoted to Lord Shiva in Kumaun.

Other gods and goddesses widely worshipped and revered in Kumaun are: Durga, Vishnu, Ganesh, Saraswati and Brahma.

Devtas: Devtas are deities who are revered and worshipped in this hilly region. In the past, they may have been mortals who were later accorded the status of deities.

a) Devtas may have been kings and rulers in the middle and early ages who solved the physical, material and spiritual problems of the people and came to their aid at times of calamity such as drought or floods.
b) Heroes of the byegone ages.
c) Mortals connected with ancient folk lore.

Devtas and devis are worshipped in each village or locality. Most village have their own devta or devi who is worshipped by all. Many families worship a *kul devta* or family devta. It is this deity that is believed to manifest in the body of the person.

Some devtas worshipped in various parts of Kumaun are:

1. *Haru*: This deity is an embodiment of the spirit of the legendary king Harishchandra who renounced his kingdom to become an asthetic. The temple of this devta is usually roofless with havan kunds or fire places. There is a trident or trishul in the centre of these kunds. The devotees smear their foreheads with ash from the havan kund for seeking blessings from the deity.

2. *Shaim*: This is the companion devta of Haru and both the temples are usually found together.

3. *Airy*: This devta is the Kumauni counterpart of the shani deity worshipped in the plains. He is taken in procession in a palanquin.

4. *Goll or Gwall*: He is the most popular deity whose temples are found all over the Kumaun hills. The temples are decorated with strips of red cloth. This deity is widely worshipped during the Navratras.

5. *Gangnath*: He is a popular deity with the lower castes and the downtrodden.

6. *Bholnath*: Another popular and widely worshipped deity. He is revered on all happy occasions and festivals.

7. *Kalbisht*: He is the spirit of a boy who was duped and murdered by his relatives. He is believed to be an accomplished flute player who grants favours to his devotees.

8. *Chaumun*: This deity is associated with animals and is understood to be their protector.

9. *Bhumiya-Khitarpal*: This deity is supposed to guard the borders of villages and human settlements.

SONGS AND EPICS

Kumauni is rich in folk dances and songs which are sung by the local people on auspicious and festive occasions, at times of sowing and harvesting in the village courtyard in the evenings.

Folk songs are also locally known as muktaks. They may be of three types or categories, viz.

a) Sung and associated with customs and traditions
b) Sung and associated with seasons and festivals
c) Religious folk songs.

Another category may be of folk songs that are sung for relaxation or recreation.

Dance songs are sung while dancing. They include jhvoar, jhora, chaanchair and chhapeti. These songs are sung in ones, twos or in groups.

There is a rich tradition of folk epics in Kumaun. They depict the following imaginary or real situations:

a) Life in this mountainous region
b) Deeds of brave people who laid down their lives in combat and difficult situations
c) Occurrences of natural calamities
d) Religious epics
e) Historical happenings in the past.

The main epics of Kumaun are given below:

1. *Malu Sahi*: A popular folk epic of Kumaun which depicts the love story of Rajuli the daughter of Sunpati Shauka and Malu Sahi, the king of Bairath.

2. *Ramaul*: A famous folk epic, dealing with the exploits of three

Ramaul youths, viz. Gangu, Sidua and Bidua. They bravely fought the Rohillas of the terai who invaded the hills of Ramulighat.

3. *Bharau and Kataku*: These are two folk epics which deal with the deeds of brave young men.

4. *Jaagars*: These are religious epics or tales which are sung in the rural areas of Kumaun particularly in times of calamities and distress.

CUSTOMS

Kumaunis have many unique customs, many of which resemble the customs of people in other parts of the country.

Naatak

The birth of a child is marked by the naatak ceremony in which the new born child and mother are purified by worshipping the gods and the family or kul devta. This is usually done on the eleventh day when the child is given a name according to his or her horoscope.

Marriage or Bya

This ceremony resembles the Hindu marriage. The salient features of marriage or bya in Kumaun are:

- —the bride is decked in typical Kumauni ornaments and garments. This includes the nose ring, bangles, ghaghra or skirt.
- —the groom comes in a small procession with his relatives and elders of his village.
- —the anchals or shawls of the couple are tied together before a sacred fire.
- —the high caste people also perform the ceremony of kanyadaan or gifting their daughter in marriage.

Sutak

This is the general name given to the ceremonies held after the death of a person. Sutak is the period of mourning which lasts for 10 to 13 days. The head of the person who has performed the last rites is shaven and he sits in mourning for this period. He sleeps on the floor and eats boiled food. In some cases the whole family abstains from eating fried and spicy foods. The purification ceremony is held after the period of sutak is over and the near and dear ones are fed on this occasion.

FESTIVALS AND FAIRS

Many festivals and fairs are celebrated in different parts of Kumaun. The salient features of these fairs and festivals are outlined below:

a) They represent the social instincts of the people of Kumaun which have been passed on from one generation to the other.

b) Experiences of living in a difficult terrain and under very severe climatic conditions have evolved many of these festivals and fairs.

c) The closely knit Kumauni society is greatly influenced by the traditional festivals and fairs.

d) Each season has its own festivals or fairs to give expression to the feelings of the people during that particular season.

e) Important agricultural operations such as sowing and harvest too are marked by festivals and fairs.

f) These are also connected with important religious and historical events.

The main festivals and fairs of Kumaun have been discussed below:

Basaht Panchami

This is one of the main festivals of Kumaun. It marks the beginning of the spring season after the long severe winter. The festival of Basant Panchami heralds the beginning of the warm climate. The people rejoice as the day becomes longer and there is revival of life in the trees of the forest. This festival also marks the harvest season for the wheat crop.

The salient features of Basant Panchami are:

a) This festival is usually held in the middle of February.

b) Saraswati, the goddess of learning and Basant (spring) are worshipped on this day.

c) The people wear yellow garments or atleast one yellow coloured garment. Women and children may put on yellow scarves while menfolk put on yellow coloured caps.

Phool devi

This is another festival connected with nature. It is celebrated on the first day of the Hindu year that is at the beginning of the month of

Chaitra. As a matter of fact, celebrations and festivities are spread over the entire month. The main features of this festival are—

a) The entrance to the house in cleaned every day and is strewn with fresh flowers.

b) Young boys and girls collect flowers from the forests and gardens; mix them with rice, turmeric and vermilion and decorate the thresholds of their houses.

c) Prayers are chanted each morning.

d) In the evening a special sweet dish is prepared and eaten.

e) During this month the lower caste people go around singing folk songs and collect money.

Hariyala

This festival is celebrated at the onset of the rainy or monsoon season. The salient features of this festival are:

a) Hariyala falls on the first day of the month of Shravan.

b) About ten days before this day, 5 or 7 kinds of grains are sown in pots by the head of the family amdist the chanting of hymns.

c) These grains sprout and the seedlings are put on the headgear on Hariyala day.

Bhitoli

This festival is held on the first day of the month of Chaitra. The brother takes presents to his married sister living in her hunsband's place.

Wallgiya

The festival of Wallgiya is celebrated on the first day of the month of Bhado in the middle of August. On this day the weaker sections are offered presents in the form of milk and milk products by the richer people. The exchange of gifts may also be between relatives or employer and employee.

Khatarua

The festival of Khatarua is celebrated on the first day of the month of Asad in the middle of September. Wood and hay are heaped at the cross roads and a bonfire is lit after dark. There is dancing and singing around this fire.

Ghughutia

This festival is held during the peak winter season in the middle of January. Birds are offered food on this day.

Fairs

Hundreds of fairs are held in different parts of Kumaun. The largest number of fairs are perhaps held in Almora district. Fairs may be of the following categories:

a) Business fairs in which there is trade or exchange of different products.

b) Religious fairs.

c) Socio-cultural fairs in which there is dancing and merry making.

d) Fairs of historical and regional significance.

ART

Traditional folk art is well developed in Kumaun. Important aspects of this are outlined below:

1. Temple art consists of paintings and sculpture on temples and other religious places. This is a very ancient form of art.

2. Art in the form of paintings on ancient rock shelters is found in a number of places in Kumaun, e.g. near Barechina in Almora district on the right bank of the Suyal river.

3. Aipan is the local name of the folk art of Kumauni women. This art is painted on the floor and walls of their houses on important occasions.

4. The floor of the places of worship is decorated with special patterns which may be a diagrammatic representation of the deity.

5. Patas are special paintings on the walls of a house or human dwelling. The subject or theme may be gods and goddesses.

6. Wall paintings may also be made of red ochre with motifs drawn with fingers using rice paste.

7. Rangwali is a unique colourful ornamentation of cloth or garments of common use.

10
Economy

Kumaun has a complex and varied economic set up. In the past, the people were solely dependant on agriculture. Now this trend is slowly changing even though agriculture and allied activities continue to be the mainstay of the economy.

1. Agriculture

Agriculture is by far the main source of livelihood for the people of Kumaun with nearly 70% of the total population dependent on agriculture, horticulture and animal husbandry.

Land Use

Kumaun has a total land area of about 21,743 sq. km of which 20.1% is under agriculture. The following table shows the land use pattern in Kumaun (U.P. Govt. Statistics, 1985):

Particulars	Area (in hectares)	% of total
1. Forest area	1127492	54.35
2. Cultivable wasteland	159559	07.69
3. Non-cultivable wasteland	70925	03.42
4. Current fallow	8676	00.42
5. Other fallow	17671	00.85
6. Land use for other than agricultural use	59310	02.86
7. Grasslands	124641	06.01
8. Land under orchards	88759	04.28
9. Net sown area	417343	20.12
10. Total	2074376	100.00

The next table gives the net sown area in different parts of Kumaun. It shows that the percentage of the net sown area to the total geographical area is very high in the plains (U.P. Govt. Statistics, 1981):

District	Sub-division/Tehsil	Net sown area as % of geographical area
Pithoragarh	Champawat	12.92
	Didihat	23.96
	Dharchula	03.83
	Munsiari	03.86
	Pithoragarh	22.67
	District average	11.54
Almora	Almora	16.19
	Bageshwar	10.70
	Ranikhet	36.01
	District average	19.42
Nainital	Nainital	08.07
	Haldwani (plain)	21.32
	Kashipur (plain)	29.56
	Kichha (plain)	79.50
	Khatima (plain)	63.59
	District average	29.34

Land Capabilities

Precise and detailed land capability studies for different parts of Kumaun have not been carried out. In general, arable land in Kumaun may be grouped into the following:

1. *High grade land*: This is extremely good agricultural land that is irrigated.

2. *Medium grade land*: This is a fairly good agricultural land which is unirrigated.

3. *Low grade land*: This is a poor quality agricultural land which is used for raising coarse kharif crops.

Agricultural Workers

In the absence of alternate sources of employment, agriculture is the main sector in which a large proportion of the working force of Kumaun is engaged. This has been amply brought out in the table given below which shows the per cent of workers engaged in agriculture in different districts:

District	*% of workers engaged in agriculture*
Almora	73.0
Nainital	45.5
Pithoragarh	78.0

This high figure in Almora and Pithoragarh districts may be attributed to the following factors:

—the small size of the operational holdings
—poverty of the agriculturists and cultivators.

The percentage of workers engaged in agriculture is low in Nainital district due to the better condition of agriculturists in the terai and bhabar tract.

Irrigated Agriculture

Bulk of the agricultural land in Kumaun is rainfed though in the hills the extent of irrigated land ranges from 8 to 12% of the total cultivated area and in the terai and bhabar tract this is as high as 70%.

The table given below shows the irrigated areas of Kumaun as a per cent of the total cultivated area:

District	*Irrigated area as per cent of cultivated area*
Almora	10.81
Nainital	68.16
Pithoragarh	08.63
Total	40.06

This figure is fairly high for Nainital district due to the extensive irrigation facilities available in the plains (mainly terai and bhabar tract). A large proportion of the irrigated land is confined to the valley and the lower parts of the hill slopes. The salient features of irrigation in Kumaun are:

1. Most of the irrigation is done through small channels locally known as kuls. Fields are irrigated by surface flow through the gravitational channels.

2. In the hills, the local people have constructed earthen or loose rock dams, across the flow of natural water channels and diverted the impounded water to channels for irrigating their lands.

3. In the terai and bhabar tract, a network of irrigation channels

has been laid which brings much needed water to the fields.

4. The irrigation department of the U.P. government too has constructed many canals in different parts of the Kumaun, both in the hills and in the plains.

Crops

Many cereal and non-cereal crops are raised in various parts of Kumaun. These have been discussed in brief in the following text:

1. *Paddy*: Paddy is widely cultivated in the following tracts of Kumaun: (a) in the terai and bhabar tract, (b) in the valleys of the hills, and (c) in terraces in the hills.

This crop is sown at the outset of the rainy season after rainwater or irrigation water has been impounded in the fields. It is harvested in late autumn or early winter.

2. *Wheat*: This is another popular cereal crop that is cultivated in most parts of Kumaun. Sowing is done in winter and the crop is ready for harvest in spring. Wheat ripens a little late in the high altitude areas due to the relatively low temperature.

3. *Maize*: Maize is popularly cultivated in place of paddy in the middle and higher hills and in the valleys.

4. *Pulses*: Many pulses are raised in different parts of Kumaun. These are listed below:

a) *Rabi season*—Chana, masur and peas.

b) *Kharif season*—Soyabean and motha are two new pulses which are being raised in Kumaun.

5. *Sugarcane*: Sugarcane is the most lucrative cash crop of the terai and bhabar tract of Nainital district. Three factors aid in a high sugarcane production in this tract: (a) advanced technical inputs provided by the G.B. Pant Agriculture University at Pantnagar, (b) fine alluvial soil, and (c) facilities for irrigation.

The total area under sugarcane in Kumaun is about 50,000 hectares.

6. *Vegetables*: Farmers in Kumaun have taken to raising vegetables as cash crops. This includes tomatoes, cucumbers, cauliflowers, cabbages, beans, pumpkins, brinjals etc. Potato is an important cash crop grown in the hills. Over 4,000 hectares is under potatoes and the average yield is about 185 quintals per hectare.

The growth of cropwise physical productivity in Kumaun (quintals per hectare) is given in the following text (after U.P. Govt. Stat., 1982):

Crop	Year		
	1960-61	1976-77	1980-81
Paddy	11.74	14.13	15.90
Maize	08.83	11.94	12.50
Madua	04.72	09.93	10.00
Wheat	08.90	08.99	13.00
Barley	07.39	06.69	10.50
Gram	05.84	09.16	09.40
Peas	09.75	09.24	09.50
Arhar	10.90	20.03	20.60
Masur	05.52	06.36	06.60

Crop Rotations

Crop rotation or succession of crops depends on the edaphic conditions, climate, irrigation available and financial resources of the cultivators. In Kumaun crop rotation is largely practised in the following three categories of land:

1. Lands which are fully irrigated in the valley upto an elevation of about 900 mts.

2. Lands which are partly irrigated in the valleys upto an elevation of about 1200 mts.

3. Lands in the uplands that are not irrigated lying above an elevation of about 1200 mts.

The common types of crop rotations practised in different parts of Kumaun are listed below:

Category	*Crop rotation*
Irrigated Land (upto 900 mts)	Wheat-paddy-wheat-paddy
	Wheat-maize-potato-maize-potato
	Wheat-paddy-potato-maize-potato
	Potato-maize-potato-wheat-paddy
	Barley-maize-paddy-barley-maize-barley
	Lentil-maize-lentil-maize
	Mustard-mandua-mustard-mandua
	Potato-paddy-lentil-maize.
Unirrigated Land (upto 1200 mts)	Fallow-paddy-wheat-mandua
	Wheat-mandua-fallow-paddy-wheat
	Wheat-mandua-lentil-fallow
	Barley-mandua-barley-mandua

	Fallow-mandua-barley-mandua
Rainfed (unirrigated)	Fallow-potato-wheat-mandua-fallow
(upto 1200 mts)	Wheat-mandua-fallow-paddy-wheat with bhatt
	Fallow-paddy-wheat-mandua-fallow-bhatt
	Fallow-potato-wheat-soyabean-fallow
	Wheat-mandua with bhatt and urad-fallow-potato-wheat.

Land Hondings

Land holdings of different sizes are owned by the people in various parts of Kumaun. This has been brought out in the table given below:

Size group (in hectares)	Per cent of total
below 0.5	10.2
0.5 to 1.0	13.6
1.0 to 2.0	22.5
2.0 to 3.0	15.5
3.0 to 4.0	09.8
4.0 to 5.0	06.8
5.0 to 10.0	13.6
Over 10.0	08.0
Total	100.0

The inferences listed below may be drawn from this table:

a) The proportion of holdings less than one hectare in size is over 70.

b) Nearly 16% of the holdings have a size varying from 1 to 2 hectares.

c) About 92% of the holdings have a size less than 3 hectares.

d) Only 8% of the holdings have an extent of more than 10 hectares.

Income

As has already been stated in the previous text most people of Kumaun, particularly those living in the rural areas depend on agriculture and allied activities for their livelihood. This is known as agricultural income.

The per capita income in Kumaun varies from area to area. It is high in the terai and bhabar tract where agricultural income is more. In the recent years there has been an increase in both agricultural and per capita income due to the availability of advanced technology and introduction of cash cropping, horticulture and dairy farming.

The household farm income and per capita income of Kumaun in 1880 and 1980 at 1988 price levels are given in the following table (Pande, 1988):

Income source	Per Household annual income (in Rs.)	
	1980	1988
Food crops	609.03	2009.00
Other crops	109.15	317.00
Fruits	127.05	568.00
Vegetables	92.19	990.00
Animal husbandry & dairying	875.54	107.00
Farm by-products	31.09	707.00
Farm miscellaneous	–	–
Total agricultural income	1844.05	5018.00
Non-agricultural income	1214.00	2300.00
Total income	3058.00	7324.00
Per capita income	511.00	1465.00

Problems of Agriculture

The following are the main agriculture problems of the Kumaun region:

1. *Fragmented and scattered land holdings*: The land holdings in Kumaun are largely fragmented and scattered. The average size of holdings is less than 1 hectare though holdings of over 10 hectares are also owned by the people, particularly in the terai and bhabar tract.

Another related problems is the absence of adequate consolidation particularly in the middle and upper hills. When land is passed on from one generation to the other it is fragmented. Consolidation of an individuals holding is necessary for higher agricultural yield. In the process of fragmentation, some land holdings become so small that it is no longer economically viable to keep them under cultivation.

2. *Low yields*: Many lands in the middle and higher hills are only capable of producing low yields. This is due to:

a) Agriculture has been extended to marginal and sub-marginal lands.

b) Lands with more than 30 per cent slope have been brought under cultivation.

c) Lack of technical inputs like improved seeds and fertilizers.

d) Poor soil status.

e) Lack of irrigation facilities.

3. *Credit*: Credit facilities are lacking in many parts of Kumaun except in the terai and bhabar tract and the valley areas. In the past, money-lenders had exploited the poor farmers to a great extent.

4. *Market*: Adequate marketing facilities are not available for agricultural produce. This is a major agricultural problem:

a) Markets if any are far from the production centres.

b) Products like vegetables may deteriorate before they reach the markets of the plains.

c) Middle men in the market make huge profits at the cost of the producer or farmer.

d) The producer or farmer is forced to sell his product at whatever price fixed by the middlemen.

5. *Irrigation*: Irrigation facilities are virtually non-existent in the middle and upper hills of Kumaun. This has an adverse effect on crop yields.

2. Horticulture

Horticulture or fruit production is slowly become popular in many parts of Kumaun. This has an advantage *vis-a-vis* agriculture and is also conducive to scientific land use. The fruits raised in different altitudinal zones of Kumaun are:

Zone	*Fruit*
Low and middle hills (upto 1500 mts)	tropical and sub-tropical fruits like guava, citrus, lemon, mango, litchi, strawberry, peach, plum and apricot.
High hills (1500 to 2000 mts)	apple, pear, peach, plum, apricot, and walnut.
Very high hills (over 2200 mts)	walnut and other dry fruits.

Fruits are grown on orchards or trees planted around homesteads and farmlands. The produce is usually collected manually and transported to the market or consumption centres. In some areas, fruits are processed into jams and concentrates which are then marketed.

Problems of Horticulture

The main problems of horticulture in Kumaun have been discussed below:

1. Non-availability of high quality planting material and seeds are big constraints in the production of fruits in Kumaun.

2. Orchards cannot be raised on holdings which are scattered and fragmented.

3. The marketing of fruits is another major problem. The margin of profit of middlemen is very high and the growers are forced to sell their produce to them at low contractual prices. In the same manner the net margins of retailers and commission agents are also high. Thus, bulk of the profit goes to the middlemen instead of occurring to the grower.

4. There is virtually no facility for institutional credit to the grower.

5. Little facilities are available in Kumaun for the processing of fruits particularly as a cottage industry.

6. The transport facilities cannot meet the pressure of the heavy seasonal production of fruit and the need for perishable commodities arriving at the market on time.

7. Packing of fruits in another major problem and they may be damaged as a result during transport.

3. Livestock/Animal Husbandry

Livestock enterprises or animal husbandry involves raising cattle, poultry, sheep, goats and pigs. This is an important activity. It is a source of income generating employment. It is a method of building stocks of nutrition. Organic matter and plant nutrients can be recycled back to the cultivated land through livestock in the silvipastoral agricultural system. Livestock can be raised on a relatively smaller area than agricultural or horticultural enterprises.

Herd Size

This is the most important parameter to judge livestock/animal husbandry enterprises in Kumaun. The average composition of livestock for farm families is given below:

Avg. no. & type of cattle	Purpose/activity
0.8 to 1.5 cows	Milk and progeny of bullocks
0.5 to 1.5 buffaloes	Milk
0.9 to 1.2 bullocks	Draft power
0.3 to 0.5 goats	Meat
1.3 to 2.2 cattle calves	Replacement
0.5 to 1.0 buffalo calves	Replacement

Milk production: Hill farmers need cattle for producing milk and for milk products like ghee and butter. This helps them to generate additional income as the surplus milk and milk products are sold by them.

In the lower hills and terai and bhabar tract, many people have taken to dairy farming. The following factors have contributed to this:

a) Easy availability of fodder.

b) Large demand for milk and milk products particularly in the urban and semi-urban centres like Haldwani, Ramnagar and Nainital.

c) Availability of high milk producing cattle.

d) Availability of support facilities like veterinary services particularly from the Agriculture University at Pantnagar.

Meat production: Sheep, goats and pigs are reared for sale as live meat. This is done in many parts of the Kumaun region. Migratory tribes too sell their sheep and goats for meat.

Wool: In the high and very high hills, people supplement their income by shearing sheep and goats and selling wool either as a raw material or in the form of finished products like shawls.

In the recent past, rearing of angora rabbits for their wool has been introduced in many parts of Kumaun. This is a profitable venture though a higher level of technology is needed for it. The animals are fed in the enclosures or sheds where they are kept and there is no danger to the forest ecosystem.

Poultry: Poultry is reared for meat and eggs which are sold on a commercial basis. Small to medium size poultry farms have been established by farmers and entrepreneurs in the terai and bhabar tract. They cater to the needs of the urban and semi-urban centres.

4. Industry

Even though agriculture and related activities form the main source of livelihood for the people of Kumaun. Other economic activities too account for a part of the gross product of this region.

Traditional Industries

The traditional industries of Kumaun consist of cottage and small scale units, viz:

a) Wool, shawls, woollen garments and carpets
b) Walking sticks and wooden handicarfts
c) Furniture and sports goods
d) Baskets and packing cases.

Requisites and Constraints for Industrialization

The main requisites for the industrial development of a region are discussed below:

a) *Protection*: Adequately protecting the resource base in totality, viz. soil, forest, water and other natural resources.

b) *Regeneration*: Regeneration of resources leading to the restoration of the forest resource, fodder resource, full utilization of the agricultural potential, effective use and management of forest resources, optimum use of fodder resources.

c) *Production*: In this process the natural resources are developed with human skill in order to enhance the productivity of the economic unit and maximise local employment and income.

The main constraints for industrialization in Kumaun are listed below:

a) Inadequate facilities like roads and communications
b) Absence of a market structure
c) Lack of adequate credit facilities
d) Lack of institutional back up
e) Relatively higher per unit cost of building basic infrastructure, social services and community facilities.

Working Population and Income

Being a hilly region, Kumaun has its peculiarities with respect to the working population. Forests which cover a substantial part of the total geographical area account for only about 12 to 15% of the regional income of Kumaun.

The distribution of the working population in different vocations in Kumaun is shown in the following table:

Working population and household industries in Kumaun, 1981 (after Dhar and Sastri, 1988)

Sector	Almora	Nainital	Pithoragarh
Percentage of workers			
Agriculture	61.80	42.66	71.36
Agriculture labour	01.92	08.52	00.03
Household industries	03.40	03.23	02.83
Others	32.88	35.56	24.78
Percentage of women workers			
Agriculture	94.84	62.91	95.76
Agriculture labour	01.67	20.20	00.46
Household industries	00.67	01.79	01.87
Others	02.82	05.10	01.71

The level of economic and industrial development of a region like Kumaun may be judged by the following indicators viz:

a) The way in which the society is organized for the purposes of final consumption.

b) The efficiency of that organization is judged in terms of volume of goods produced, consumed and the magnitude of the services required.

c) Structural and altitudinal aspects of the pattern of socio-economic organisation.

d) The willingness to take economic risks

e) The accumulated economic infrastructure

f) Social beliefs and attitudes towards work

g) Occupational structure of the regional economy.

The salient features of the working population and income levels/ distribution have been brought out in the following text:

1. The total manpower engaged in agriculture in Nainital district is lower than the other two districts of Kumaun. There is also more use of inputs like fertilizers and irrigation. As a result the overall production in this district is higher. In Nainital district there has been a gradual development of the secondary and tertiary sectors.

2. A new category of marginal workers have emerged in Kumaun, particularly in Nainital and Almora districts due to the development of secondary and tertiary sectors. This occupational shift is more towards the tertiary sector.

3. Illiterates are engaged mainly in the agriculture sector while skilled, semi-skilled and educated workers are absorbed in sectors like industry.

4. The occupational shift shows an increasing absorption of females in the primary sector with more absorption of the male labour force in the secondary and tertiary sectors.

Present Status of Industries

The present status or condition of industries in Kumaun has been discussed in the following text:

1. *Traditional industries*: The two most important traditional industries in Kumaun are weaving of ringal for making various items and weaving of wool based goods. The former is largely undertaken in areas where ringal is available within a short distance. At present this activity is more concentrated in and around consumption centres, viz. Chaukhutia, Takula, Dhauladevi, Bageshwar, Kapot, Hawalbag, Dwarahat, Tarikhet, Som, Karmi, Jhaulauj, Faldua, Rudoli, Chintoli and Jogeshwar. The workers make baskets, grain-bins and articles of domestic use from ringal. They also make ropes, twines and sacks from locally available species. However some factors are responsible for the decline of this industry such as: (a) non-availability and rising cost of raw material, (b) unremunerative prices, and (c) intense competition offered by more durable, water and moisture proofs substitutes now available in the market at readily affordable prices.

Weaving of wool-based products such as shawls and carpets is a local industry in the higher hills, viz. Talla Johar, Malla Johar and Darma area. Wool from goats and sheep is used for the purpose. This is an important economic activity of these areas. However, the wool weaving industry has declined due to the causes listed below:

a) Sealing of the border with Tibet which has virtually stopped cross-border trade.

b) The availability of skilled labour has become scarce with diversification of employment avenues.

c) Decline in availability of raw material.

d) Carpet weaving has become uneconomical in many areas with constraints of raw material supply and costs and non-realisation of adequate prices.

e) Obsolete technology and design and limited colour patterns used by the carpet weavers of Kumaun.

f) Stiff competition from artisans based at Badhoi, Agra, Mirzapur and Jaunpur.

Blacksmithy, coppersmithy and carpentry are other traditional industries of Kumaun. Of these blacksmiths and carpenters are based in most urban and semi-urban centres due to the demand for their products. Coppersmiths are more in Almora, Takula, Islona, Pagana, Bagargaon, Karaipatti, Bagbazar, Mayagaon, Bageshwar and Chaukhutia. They are either full time workers or devote more than fifty per cent of their annual production time to this economic activity. In case of coppersmithy too, there has been a general decline due to stiff competition offered by products from places like Moradabad.

Other traditional industries of Kumaun are stone carving, extraction of slate for roofing, leather tanning, basket making, sawing, logging, hand spinning and mining. There is a gradual downwards trend in all these activities.

2. *Village and small industries*: There are many village and small industries in Kumaun. These include spinning, weaving, furniture making, brassware, handicrafts and sericulture. Many units are located in the terai and bhabar tract of Nainital district. These are being promoted through various schemes such as:

a) Integrated rural development programme (IRDP)
b) Development of women and children in rural areas (DWCRA)
c) Entrepreneurial development programmes
d) Single window assistance concept.

About 5,000 small and village units have been set up with the following assistance from such promotional programmes:

a) Subsidy on generating sets
b) Assistance to sick units
c) Margin money loans
d) Five per cent state capital subsidy
e) Five per cent additional subsidy to ancillary units.

The Khadi and Village Industries Commission has started many programmes to train the people in vocations like spinning and weaving of wool, bee-keeping. It also gives financial assistance to weavers cooperatives.

3. *Medium and large industries*: Many industries in the medium and large sector have been set up in Kumaun, particularly in the terai and bhabar tract viz. Kashipur, Rudrapur, Kicha, Haldwani, Ramnagar and Kathgodam. These are both under private and government management. Some of the larger units are listed below:

1. H.M.T. Watch unit, Ranibagh
2. B.H.E.L. components, Rudrapur

3. U.P. state spinning mills, Kashipur & Jaspur
4. Century paper and pulp mill, Lal Kuan
5. L.H. sugar factory,Kashipur
6. Gauraya straw card board mill, Kashipur
7. Jindal vanaspati udyog, Kashipur
8. Belwal spinning mill, Pipalsana, Ramnagar
9. Prakash tubes, Kashipur
10. Bimtal photofilms,Bhimtal
11. Indira Gandhi cooperative sugar mill, Sitarganj
12. Sahkari soyabean vanaspati, Haldwani
13. Ganesh flour mill, Khatima
14. Eastman agro mills, Lalpur, Rudrapur
15. U.P. digitals, Gorakhal
16. Almora magnesite, Jhiroli
17. Saraswati woollen mills, Ranikhet
18. Himalayan magnesite, Chandak, Pithoragarh

5. Forestry

The forests of Kumaun are under the Chief Conservator of Forests, (CCF), Kumaun, whose headquarters are located at Nainital. He functions under the Chief Conservator of Forests (Hills) and Principal Chief Conservator of Forests, both of whom are at Lucknow.

Under the CCF (Kumaun) are three circles each headed by a conservator of forests, viz. Kumaun, western and timber supply circles. These in turn are made up of territorial forest divisions each of which is headed by a divisional forest officer. The boundary of a forest division does not coincide with that of a district.

Little is known of the history of forest management in Kumaun prior to the advent of the British in 1815. Deodar trees were extensively planted around temples, e.g. Jageshwar. Forests were perhaps considered to be an inexhaustive resource in terms of fuelwood, fodder and timber for the construction of houses. However, the process of degradation of forests had already begun and the British introduced scientific management of the rich forest resources of Kumaun.

Three types of forests have been constituted in Kumaun, viz.

a) Reserved forests
b) Civil or protected forests
c) Panchayat forests.

The forests area of different districts of Kumaun is given in the following table:

District	Forest area (in sq. kms.)	Geographical area (in sq. kms)
Almora	3944.32	5385
Nainital	4026.40	6794
Pithoragarh	3302.89	8856

Forest Utilization

Forests of Kumaun are utilized for the following purposes both on a commercial and local basis.

1. *Timber*: Timber from the forest is used for structural purposes, viz. for the construction of houses, bridges, for furniture making, plywood, veneers, paper, pulp, matchwood, electric and telephone lines.

2. *Fuelwood*: This is one of the most important uses of forests in Kumaun. Bulk of the energy in the rural areas comes from forests in the form of fuelwood. The local people remove dry twigs and branches and lop green trees whose branches are dried and used as fuel. This is used both for heating and cooking purposes.

3. *Fodder*: Fodder is another product that is widely obtained from the forests of Kumaun. This is in the form of green leaves, succulent twigs, shoots and grasses that are lopped for stall feeding cattle. Domestic animals may also be taken to the forests to browse and graze all day long. Thus, in the absence of enough land for raising fodder crops, except in the terai and bhabar tract, the fodder for bulk of the cattle population of Kumaun comes from the forest.

4. *Other products*: Other important products obtained from the forests of kumaun are:

a) Food and spices
b) Medicines and drugs
c) Resin, mainly the resin of chir pine
d) Fibres and flosses.

6. Tourism

Tourism is an important economic activity in many parts of Kumaun. Important places of tourist importance in Kumaun are listed below:

Almora district: Almora, Ranikhet, Jageshwar, Binsar, Baijnath, Kausani, Chaukhutia and Pindari glacier.

Nainital district: Nainital, Bhimtal, Corbett National Park, Ghorakhal, Kainchi and Nanakmatta.

Pithoragarh district: Pithoragarh, Lohaghat, Dharchula, Munsiari and Milam glacier.

Five types of categories of tourists visit Kumaun. These are:

a) Higher class tourist who books and plans his trip well in advance. He wants the best facilities.

b) Upper middle class tourist who wants facilities that are good but not expensive.

c) Budget tourist with a limited income and limited capacity to spend.

d) Freelance tourist consisting mainly of youth and non-earning groups.

e) Religious tourist or pilgrim.

f) Mountaineer and trekker.

Salient Features

The salient features of tourism in Kumaun have been discussed in the following text:

1. Nainital is perhaps the most popular tourist resort of Kumaun. The lake and its surrounds are the most popular attraction of this town. Other similar lake resorts near Nainital include Bhimtal. The concentration of a large number of tourists at Nainital has led to the ecological degradation of this lake.

2. There are a number of mountain resorts in Kumaun which offer an escape from the scorching heat of the plains in summer, e.g. Ranikhet, Almora and Kausani.

3. Places of adventure tourism are the Pindari and Milam glaciers.

4. Places of wildlife tourism are the Corbett national park and Binsar sanctuary.

5. Places of religious tourism in Kumaun are Jageshwar, Bageshwar, Baijnath and Someswar.

6. There are a number of places of sight seeing interest which are visited by tourists, e.g. Garbyang, Dharchula and Malla Johar.

7. Minerals

Kumaun is rich in mineral deposits. Amongst the mineral deposits which can be or are being exploited are:

1. *Dolomite*: Found in many parts of this hill region. This is an important refractory material that may be used as a liner in kilns and blast furnaces. Economically exploitable dolomite deposits in Kumaun are located at:

a) Jhulaghat-Gangolohat-Jhirauli
b) Panthsera-Rainagar-Kanda
c) Dharchula-Kapot and extending into the Pindar valley
d) Bajun-Nandhaur valley near Kundal

2. *Limestone*: Economically exploitable deposits of lime stone are found in many areas of Kumaun. These are being quarried or mined at:

a) Baste-Naulara (Ramganga valley)
b) Postala-Kulur gad area
c) South of Gangolihat to Nargol gad, Bagaur, Chaunala and Suitola
d) Chanakhan-Saulena-Adhaura (Nainital area)

These limestone deposits are being used for the manufacture of cement and slaked lime.

3. *Magnesite*: Economically viable deposits of magnesite are found in three belts, viz.

a) Chandaak-Dharigaon
b) Panthsera-Dungi-Dewallthal-Rainagar-Kanda
c) Painya-Tuper Jakh-Jakheri (Lahor valley)

These deposits are being mined by the Magnesite Mineral Ltd. and Almora Magnesite Ltd.

4. *Talc*: High grade talc is found associated with the magnesite deposits in parts of Almora and Pithoragarh districts. These deposits are in the valley of the rivers Lahor, Pungar, Ramganga and Gori. Talc is being mined on a commercial basis in a number of places.

5. *Building stones*: Stones for the construction of houses, viz. slates, sandstone and quartzite are quarried in many parts of this region.

Mineral based industrial units being set up in Kumaun are listed below:

1. Chaunala cement plant, Gangolihat, Pithoragarh
2. Matela cement plant, Almora
3. Jaurasi cement plant, Kosi valley, Nainital
4. Bhulgaon mini-cement plant, Almora

5. Polymetallic ore mining and concentration project, Askot, Pithoragarh
6. Tungsten mining concentration project, Patsal and Chaukhutia, Almora.

8. Hydel Projects

There is immense scope for the development of hydro-electric projects in Kumaun. A number of such projects are already functioning. The rivers have been dammed with the erection of small dams and barrages and the reservoir created as a result is being used to feed water for turning giant turbines which in turn produce electricity. The impounded water is also used for irrigating fields in the downstream areas by means of canals.

Such projects have been set up along the Ramganga, Kosi, Kaliganga and Sarju rivers and their tributaries. A number of projects are being constructed.

9. Rail Lines and Roads

Rail lines and roads are extremely important for the economic development of a region. Rail lines connect Kashipur-Ramnagar; Kashipur-Bazpur-Lalkuan-Khatima-Tanakpur and Kicha-Lalkuan-Haldwani-Kathgodam in the foothills of Kumaun.

A network of roads connects most places in Kumaun. Important road links of Kumaun are:

1. Rampur-Haldwani-Nainital-Almora
2. Haldwani-Lohaghat-Pithoragarh-Dharchula
3. Pithoragarh-Munsiari
4. Kashipur-Ramnagar-Ranikhet-Almora-Baijnath
5. Ramnagar-Kaladhungi-Haldwani-Tanakpur
6. Karanparyag-Gwaldam-Berinag-Thal-Dharchula
7. Chaukhutia-Dwarahat-Ranikhet-Almora.

11

Environmental Degradation

Like the adjoining tracts of the western and central Himalaya, the Kumaun hills too have been subjected to severe environmental degradation in the past few decades. This aspect has been discussed in detail in the following text:

Biotic Pressure on Forests

Forests cover a substantial part of the total geographical area of Kumaun. In the last four decades man has interfered with the forest ecosystem as more and more wood is extracted for fuel, fodder for cattle and timber for the construction of houses in the rural areas.

Overgrazing

Overgrazing is perhaps the most serious factor responsible for the ever increasing biotic pressure on Himalayan forests. Over the years with an increase in the cattle population there has been a rapid degradation of the forest environment.

The hill people rear cattle primarily for the following purposes:

1. *Milk and agricultural uses*: Cows, goats and in the lower hills buffaloes are reared for milk. There are enough number of cattle in a village to cater to its milk requirements. However, small villages have to depend on semi-nomadic and nomadic grazier communities for their milk supply. They also supply milk to towns like Nainital, Ranikhet and Almora. The hill cattle yield less milk. Thus, there is a scarcity of milk in this region, particularly in the higher tracts.

Bullocks are used for ploughing fields and for drawing carts in the terai and bhabar tracts. Thus animals are short but sturdy in structure as they have to pull the plough on terraces which may be very small in size. Animals used for ploughing are generally stall-fed while those meant for milk production feed on forests and pastures.

2. *Manure*: Animal dung has a prominent place in the agriculture of the Kumaun hills. Modern fertilizers are not very widely used except in the foothills. The animals are grazed on pastures and forests during the day and their night dung is collected for manuring fields.

In a way, forests and pastures also benefit from animal dung as the day dung, particularly of goats and sheep is seldom removed from the forest and pasture floor. Due to extreme cold conditions prevailing in most parts of Kumaun, the dung of domestic animals is collected and later on used for heating and cooking.

3. *Wool*: Semi-nomadic and nomadic tribes in Kumaun rear goats, sheeps and yaks for their hair which is used for making wool and woollen garments, blankets and carpets. Animals are specially reared for this purpose. Very often they are left to roam in the forest for several days at a stretch. This practice has gained popularity in recent years due to decrease in the numbers of the predator species.

The semi-nomadic and nomadic tribes of these hills are altitudinal migrants quite like their counterparts in other parts of the Himalaya. In summer they move with their cattle to the high altitude forests or pastures where the animals feed on the lush green meadows. They leave these areas in autumn and reach the foothills or the shelter of the valleys in winter.

Degradation

1. Gullies form in the tracts much frequented by cattle. This is more common in the Siwalik hills where gully formation is facilitated by the weak country rocks.

2. Overgrazing has an adverse effect on the pasture grasses by keeping them below the optimum height for metabolic activity.

3. Due to the selective grazing of young and succulent grasses, only coarse and poor grasses remain in the pasture.

4. Cattle hooves also damage young regeneration, seedlings and even saplings. Their hooves scour the soil, disturb its compactness and aid in accelerated erosion.

Forest Fires

Forest fires are also an important form of biotic pressure on forests and play a significant role in forest degradation.

Intentional fires: These fires are set intentionally with a specific intention to set fire to the forest.

1. Miscreants may set fire to the forest in order to damage or destroy the forest wealth.

2. Villagers or the local people may set the forest floor on fire in the summer months so as to induce a good growth of grass after the rains are over.

This may also be done for a good growth of mushrooms which is collected as a minor forest produce.

Unintentional or accidental fires: Unintentional fires may be started by picnickers, trekkers and travellers who may leave unextinguished cigarette butts, match-sticks and camp fires in the forest paticularly during the summer season.

Degradation

1. Forest fires damage the vegetation to a large extent.

2. It harms the young regeneration and burns the seedlings and saplings.

3. Forest fires completely destroy the vegetation growing on the forest floor. This is an open invitation to accelerated erosion. Rills and gullys are formed in the aftermath of forest fires.

4. The valuable fauna is also destroyed during forest fires.

Overlopping

In the evenings it is usual to see hill women-folk and children returning from the forest with a load of fuel wood or grass on their heads. As there is virtually no alternate source of energy available to the rural people in Kumaun, they have to depend on fuelwood obtained from the nearby forests for their cooking and heating requirements.

Most villagers enjoy certain rights and concessions for the collection of dry stems, barks and leaves from the neighbouring forests for their energy needs. However, dry stems available from a forest is seldom enough to meet the requirements of an average village having about 20 to 30 families. Thus the villagers extract even green stems for use as fuelwood.

Lopping is also done for stall feeding the cattle. Under the rules only the lower one-third of the crown of a tree is to be lopped. However this is seldom adhered to. The above practices lead to overlopping which has become a major problem in the lower hills where the population pressure is more than what the forests can sustain.

Degradation

1. The regeneration of the future crop is adversely affected as flowers, fruits and seeds may also be lopped. Overlopping of the crown of mature trees also reduces their potential to bear flowers, fruits and seeds.

2. The canopy is opened, thus paving the way for problems like accelerated erosion.

3. The site quality of a forest is reduced.

4. There is an adverse effect on the productivity of the trees.

Reduction of Forest Area

Before the enactment of the Forest Conservation Act in 1980, there was reduction in the forest area of Kumaun due to the activities of human beings.

1. River valley projects involve the submersion of large tracts of forest land under the reservoirs created when dams are erected across rivers.

2. In the past as the population in these hills swelled, more and more forest land was cleared and brought under agriculture and horticulture.

3. Avalanches, landslides, floods, storms, earthquakes and cloudbursts etc also lead to a reduction in the forest area of a locality or region.

4. Roads are important links between people and places. In Kumaun hundreds of kilometres of roads have been constructed. This means clearing of valuable forest land.

Environmental Problems due to Biotic Pressure on Forests

Biotic pressure on the forests of Kumaun leads to the following environmental problems:

1. *Accelerated erosion*: Degraded forest lands cause accelerated erosion. It is a well known fact that leaves break the force of the falling rain water. Negi (1982) states, "the many tiered canopy of the Himalayan forests act as a brake on the force of the rain drops. The drops hit the crown of the topmost tree, then fall onto the herbs and shrubs below. By the time the precipitation reaches the forest floor its force is greatly reduced."

As a matter of fact, in a dense forest, a considerable part of the rain water trickles down along the stems of trees, shrubs and herbs. Ghosh and Rao (1979 as cited by Negi, 1982) have arrived at the following results regarding the per cent interception of rainfall of different tree species (most of which are also found in Kumaun):

Species	*Per cent interception*	*Trees per hectare*
Sal	25.30	1678
Khair	28.50	574
Eucalyptus	11.56	1958
Chir pine	22.10	1156

In areas with little or no vegetative cover the raindrops fall directly on the forest floor. Due to its kinetic energy this drop dislodges soil particles and hence this phenomenon is termed as splash erosion. These drops combine to form channels which form figure like rills that ultimately lead to the formation of gullys and canyons. Such features are common on the dry and exposed slopes of the lower hills of Kumaun.

Vegetation helps to bind the soil together. The root with its network of countless roothairs keeps the soil particles together. Devoid of the valuable root network the topsoil is quickly washed away by the rain water. Huge chunks of unprotected soil give way to the force of the rain water and soil slumping results. Very often, unprotected soil acts as a catalyst in the formation of massive landslides.

In the higher hills, forests also provide protection against erosion by snow. The tree canopy reduces the quantity of snow falling on the ground. Melt waters from the snow covering the ground flows away in channels and causes considerable soil erosion. A thick layer of humus and undecomposed organic matter on the forest floor acts as a sponge and gives more time to the rainwater and meltwater to seep into the soil. Snow accumulates on the crowns of trees and when it melts the water trickles down along the trunk, thus reducing the risk of accelerated erosion.

As surface water flows down the mountain side every tree, shrub and herb will impede its flow and arrest the eroded material flowing away with it. The vegetative cover will also reduce the high velocity of the surface run off thereby lowering to a large extent its capacity to erode the soil.

Thus, it is clear that in the absence of a forest cover, accelerated erosion sets in. Very soon the mountain slopes are bare and devoid of a soil cover. In valley areas, the rock and soil debris brought down by the surface run-off may cover valuable arable land, roads, canals and even human settlements.

2. *Vegetative changes*: Degradation of the forest cover has caused retrogressive changes in the vegetation of many parts of Kumaun. This problem is also very prominent in the adjoining Garhwal hills. Bhatt (1980) states, "today, large parts of Uttrakhand once known for its dense forests of oaks, pine, deodar and other species now lies denuded. Along the banks of the Himalayan rivers hundreds of kilometres of land strips are now clustered with cactus alone and the whole region is gradually turning into a desert."

3. *Climatic changes*: Forest areas have a moderating effect on the climate. In summers it is cooler inside a forest than in open and exposed tracts. The bare rocky slopes heat upto the surrounding atmosphere by acting as a mirror. Trees also break the force of the winds that roar across the slopes thereby minimising erosion by wind.

4. *Lowering of the water-table*: Let us assume for a moment that the mountain slopes were made up of glass. Any precipitation on these slopes would lead to flooding of the plains. The power or capacity of soil to absorb water depends on its permeability which is maintained by the roots that act as a disintegrating agent, leaf litter and humus. On steep slopes where the soil is shallow, the forest floor litter plays a major role. It absorbs about 3 to 4 times more water than ordinary soil and by easy decomposition adds to the depth of the soil.

Less water seeping into the soil leads to the lowering of the water-table. Water is the very basis of civilization particularly in the Kumaun hills where bulk of the water needed by the people is obtained from springs and streams fed by underground water.

Now with the increase of biotic pressure on forests, the water-table has been lowered. Women and children have to walk further and further to fill their pots and cans have to be lengthened as the nearby sources of water are drying up. The discharge of many springs in the middle and lower Himalaya has considerably declined. In some tracts it is common to come across springs that have been reduced to a mere trickle or have completely dried up.

5. *Flash floods*: Flash floods have become more frequent in the Kumaun hills in the last few decades. Degradation of the forest cover in the catchments of rivers and streams due to increased biotic pressure is responsible for this. The tree less mountain slopes are unable to absorb the vast quantity of rain water that pours down from the skies in a short time and this water ends up in the rivers and streams whose channels are narrow. They overflow their banks and cause widespread

loss and misery.

6. *Slope stability*: Slopes devoid of a vegetative cover are prone to soil slumping and landslides. The roots of plants bind the soil together and thus help in slope stability particularly on very steep gradients. In the absence of a forest/vegetative cover the thin mountainous topsoil becomes susceptible to the slightest disturbance. Rain water renders the rock and soil debris highly viscous and thus it moves downwards, covering arable lands, forest areas, roads and even houses and blocking rivers and streams.

7. *Leaching*: Surface run off is responsible for the leaching away of important nutrients from the soil. In the absence of a vegetative cover the surface run off washes away the clayey components of the soil leaving the coarse and unproductive soil behind. Thus, the more the run off, the larger will be the quantity of nutrients lost from the soil.

Most forest soils in Kumaun are rich in organic matter due to the thick layer of undercomposed humus that covers the topsoil. Huge torrents of the surface run off will leach away this important constituent. On the other hand if the same water were to seep into the soil it would carry the nutrients to the layers below, thus enriching them.

8. *Soil formation*: As they grow, the roots of the soil are responsible for the disintegration of all rocks, whether it is granite, quartzite, volcanic or shale. Roots of trees and shrubs play an important role in the formation of soil. They combine with other elements such as water, frost and gravity action to produce the soil cover. They help in rock disintegration as they grow downwards, by penetrating into the small cracks and fissures, thus paving the way for water to act upon it.

In the absence of a forest cover, the process of soil formation is adversely affected.

9. *Carbon sinks and oxygen production*: Forests act as sinks for carbon dioxide and produce the all important oxygen needed for the survival of human beings. In the absence of forests this gaseous exchange process is adversely affected.

Quarrying

Kumaun is rich in minerals like limestone, dolomite, magnesite and building stones. These are being extracted, primarily by the process of quarrying or surface mining. As a matter of fact people living in these hills have been quarrying slates and other building stones for house construction since times immemorial. Old workings of copper and iron are also found in some parts of these hills.

However regular quarrying on a considerably larger scale started in the fifties of this century. This involves the following steps:

1. All trees, shrubs and herbs growing on the area to be quarried are removed. The ground flora may be set on fire.

2. The topsoil is removed either manually or with the help of earth moving machines. The debris is thrown down the hill slopes.

3. Bench or step like features are cut on the exposed slope and the mineral extracted.

4. It is transported from the quarry manually, by trucks or ropeways.

On the other hand underground mining tunnels are dug deep into the bowls of the earth and the mineral extracted from them. This method is less damaging to the environment.

Environmental Problems caused by Quarrying

Quarrying or surface mining leads to the degradation of the environment in the following manner:

1. *Loss of vegetation*: The forest/vegetative cover made up of valuable trees, shrubs and herbs is lost for ever. This causes associated problems like accelerated erosion, rill and gully formation.

2. *Loss of valuable land*: In quarrying, valuable land that could have been put to other protective uses is lost. It is difficult to reclaim abandoned quarry sites in the difficult terrain of Kumaun.

3. *Disturbance to wildlife*: Quarrying disturbs the wildlife of the area which is forced to move away to other areas.

4. *Loss of productive topsoil*: Nature takes thousands of years to produce a few centimetres of topsoil which is the base for the growth of all kinds of vegetation. When the overburden is cleared in open cast mining, this valuable topsoil is lost within the matter of a few hours. In fact it is lost for ever as nature will never be able to rebuild it.

5. *Hydrological pollution and associated problems*: a) The local water table is lowered in and around quarry sites as less water seeps into the soil.

b) Devoid of the protective cover of topsoil the exposed rocks are subjected to the full action of the falling rain water. Some chemicals like calcium carbonate may be dissolved by the rain water, thus making it unfit for human consumption; agricultural or industrial purposes. When this water containing dissolved chemicals reaches the larger streams and rivers, it may pollute them thus affecting aquatic life.

c) Water containing unwanted dissolved chemicals seeps into the rocks and finds its way into the reservoir of underground water, thereby contaminating the water which may be consumed by human beings in many villages.

5. *Mineral transport*: Roads may be constructed right upto the quarry site so that transport of the extracted mineral becomes easy. This leads to loss of vegetation and soil and causes the occurrence of landslides. There is also an adverse effect on the drainage, agriculture and water quality.

6. *Air pollution*: Quarries are also a source of air pollution. They may cause certain lung diseases like silicosis in the labourers working there.

7. *Sediment inflow*: A vast quantity of sediment flows with the run-off from the quarries. The erosive capacity of the sediment bearing surface run off is greatly increased and it leads to the formation of rills and gullies in the areas lying below the quarry site.

8. *Debris*: The debris from quarries is dumped down the hill side. It covers valuable arable land, forest land, roads, sands and even human settlements.

9. *Visual pollution*: Quarries are ugly looking repulsive scars on the mountain sides which are not a pleasant sight for eye.

Landslides

Landslides or slope failure is a major environmental problem in Kumaun. They cause loss of crores of rupees on account of man and vehicle hours lost, loss of property, settlements, houses, fields, forests, roads, goods delayed and damaged and cost of road maintenance and removal. Virtually no part of this hilly region is free from the problem of landslides.

Factors Causing Landslides

The following factors are responsible for landslides or slope failure indifferent parts of Kumaun.

1. *Changes in slope gradient*: A sharp increase in the gradient of the natural slope brings about abnormal changes in the internal stress of the rock and soil mass, thereby disturbing the conditions of equilib-

rium by an increase in the shear stress. This may be brought about by artificial or natural factors such as undercutting of the base of a slope by flowing water or by excavation. In some cases the angle of the slope may be abruptly changed due to tectonic processes such as subsidence or uplift.

2. *Changes in slope height*: This may be brought about by vertical erosion or excavation. The deepening of a river valley relieves lateral stresses which in turn loosens the rocks and soil resting on the slope and forms fissures parallel to the slope surface. Rain water thus penetrates through these fissures and acts as a catalyst in the occurrence of a landslide.

3. *Overloading*: Overloading by embankments, fills and soil heaps may lead to the formation of a landslide. This produces an increase in the shear stress and also an increase in the pore pressure of clayey soils. Thus the shear strength is decreased. A quicker pace of overloading is more dangerous.

4. *Changes in soil and rock water content*: Changes in the soil and rock water content may lead to the occurrence of landslides. This is more frequent during periods of very high rainfall (during the monsoons) when rain water percolates into the joints and fissures and produces a hydrostatic pressure. A pore pressure is thus generated in the soil and this causes a decrease in the shear resistance.

Difference in the electric potential when two beds are in contact along a sliding plane may also lead to the occurrence of a landslide. In clayey rocks the deleterious effect of rainwater is more when rainfall occurs after a fairly long dry spell. The water is able to rapidly percolate deep into the fissures as the clayey soils are desicated and landslides result.

5. *Vegetative cover*: It is a well known fact the network of roots of trees, shrubs and herbs help to bind the soil together. They also help to break the force of the falling raindrops. Another function of the vegetative cover is to help to stabilise soil slope.

In the absence of a vegetative cover or when the forests covering a slope are degraded, this protective role is adversely affected and the slope fails, thereby causing thousands of tons of rock and soil debris to roll down the mountain slopes.

6. *Weathering*: Mechanical and chemical weathering disturbs the cohesion of the rocks. Chemical alterations such as hydration and cation exchange in clays may act as a catalyst in the occurrence of landslides. Mechanical weathering may also upset the stability of

slopes and lead to the occurrence of a landslide.

7. *Tectonic causes*: Faults or thrusts are planes along which two blocks of the earth's crust move due to tectonic movements or pressures. They produce wide zones of weakness or dislocation. Very often the slope in such zones fails leading to the occurrence of landslides.

8. *Road construction*: The construction of roads in a particular area may upset the stability of the natural slopes, thereby causing landslides.

Landslides in Pithoragarh Area

Pithoragarh district occupies the north-eastern part of the Kumaun hills. Landslides have been occurring in the Tawaghat-Kalika Devi-Dharchula tract of this district for quite sometime. The tract falls within the catchment of the Kaliganga river that has carved a deep V-shaped valley. The rocks consist of the central crystallines and the lesser Himalayan sequence. The main central thrust which is a deep seated tectonic dislocation in the earth's crust has formed a sharp angle in this tract, thereby weakening the country rocks.

On August 15, 1977 after several days of incessant showers there occurred a giant landslide near Tawaghat. On this day, thousands of tons of debris hurtled down the mountain side whose slope varies from 40° to 70° or even vertical at places. Two eye witnesses, Gagan Singh and Jai Singh, chowkidars of a rest house at Tawaghat were spell bound as they saw gigantic rocks hurtle down the mountain slopes at a great speed. As a result of this slide, life was disrupted in an area of about 25 sq. kms. More than 1.5 kms of road in various sections was affected. Forty four human lives were lost and a still larger number injured. Over 50 houses were damaged and many cattle perished.

In 1977, another massive landslide occurred between Palpala and Khela village in the same area in which a number of human and animal casualties occurred.

Similar landslides occur each year in the rainy season in different parts of Kumaun.

Environmental Problems due to Landslides

Landslides are responsible for a number of environmental related problems in Kumaun. These have been discussed in brief in the following lines:

1. They cause a colossal loss of man and vehicle hours in terms of road blocks and traffic disruptions.

2. Human beings and animals may either perish or be injured when landslide debris comes rolling down the mountain slopes.

3. They may damage or destroy: (a) buildings, villages or other human habitation, (b) arable, pasture or forest lands, (c) roads and bridges, (d) dams and canals, (e) reservoirs and lakes, and (f) pipe lines, telephone lines and power lines.

4. Landslides may temporarily block the flow of a river or lake thereby creating a lake. When this dam bursts a vast quantity of water is released into the stream or river channel which floods its banks and leads to further damage.

Degradation of Lakes

There are many lakes around Nainital including the Nainital and Bhimtal lakes, which are the largest amongst these. Each year tourists flock to these lakes in the summer season to escape from the scorching heat of the plains. Townships have sprung up around these water bodies. Of these, Nainital is by far the largest followed by Bhimtal. Over the years, Nainital and Bhimtal have become very popular with tourists. This has increased the biotic pressure on the lakes, particularly Nainital.

The ecological degradation of these water bodies has been caused by the following:

1. *Sewage disposal*: The sewage including human excreta is emptied into these water bodies. The quantity of sewage received by the lakes is considerably more in summer due to the increase in the number of tourists.

2. *Accelerated erosion*: Being tourist resorts, these townships have experienced a rapid growth. This has disturbed the vegetative cover in the catchment area of the lakes, thereby leading to an increase in the quantity of silt that flows into the lakes.

3. *Boating and sailing*: Boating and sailing are popular recreational activities with tourists. Boats have disturbed the ecological balance of these water bodies.

Environmental Problems

The environmental problems caused by the degradation of the Nainital, Bhimtal and other lakes in this region has been outlined in the following text:

a) *Dissolved oxygen*: The level of dissolved oxygen in the surface layers of the waters of Nainital and Bhimtal lakes is around 100% saturation. In Nainital the level of dissolved oxygen during the spring season is less than 5 mg/litre. During the summer months which is the peak tourist season this falls to below 2.5 mg/litre. This condition is dangerous for fresh water fishes.

There are cold water carps in Nainital lake which are sensitive to low oxygen concentration. Thus a substantial number of these fishes die in winter when the level of dissolved oxygen falls below the critical limit, i.e. 0.9 mg/litre. Deficiency of oxygen in the hypolimnion water and the entire body of water in winter in the Nainital lake can be related to the decomposition of accumulated organic matter in the lake bottom and decaying of large quantities of algae which blooms frequently then die and finally sink to the bottom.

b) *Nitrogen and phosphorous level*: The level of nitrogen and phosphorous have a significant bearing on the biological activities of aquatic systems. In Nainital lake the NO_3-N reaches a peak value of about 500 mg/litre in autumn and decreases to a lowest value of about 75 mg/litre in spring. In Bhimtal lake this varies from 0.0 mg/litre in spring to 40 mg/litre in autumn and early winter. There is also a wide variation in the levels of phosphorous.

On basis of nitrogen and phosphorous levels, the Nainital lake can be said to be in a fairly advanced level of eutrophication while the Bhimtal lake is in the pre-eutrophication stage.

c) *pH, free CO_2 and organic matter*: At the surface, the pH of both Nainital and Bhimtal lakes is between 5 to 7. However, water samples from a depth of 3 mts show a slightly acidic level in the case of Nainital lake.

The free CO_2 in Nainital varies from 27 mg/litre in winter to 21 mg/litre in the rainy season. In Bhimtal this level is about 10 mg/litre in winter. Organic matter in Nainital is between 1000 to 1500 p.p.m.

d) *Harmful substances*: Due to the disposal of sewage and other waste products in the Nainital lake, there is formation of ammonia, hydrogen sulphide, methane and other toxic substances at the bottom of the lake. This has an adverse effect on the physiology of the biotic organism in the lake.

e) *Biological organisms*: Pollution of the lake waters has adversely affected the biological organisms living in them.

Fish: A large variety of fishes are found in the lake waters. This includes trout, mahaseer, mirror, crap, asela, rohu, catla and mrigal. Pollution of the lake waters has resulted in two major physiological changes in fish life.

a) There has been an increase in the fish mortality rate during the past three decades. Each year about 40 to 50 kg of fish perish in the Nainital lake. There has also been a decline in the fish catch from this lake.

b) The spawning and breeding grounds of the carp fish have been destroyed by sewage and silt. Fishes breed in shallow waters near the periphery. It is along the banks of the lake that the ecology has been distrubed to a considerable extent.

Water Molds: Water molds are saprophytic in nature and need organic matter to survive. An abundance of water molds in the waters suggests a high organic matter content due to more sewage inflow. Concentration of organic matter in the lake waters affects its drinking water quality as different pathogenic micro-organisms are directly or indirectly associated with organic products and are responsible for a number of water borne diseases in human beings and animals.

Zooplanktons: All the lakes of Nainital area contain zooplanktons. Amongst them species like *Cephalodella*, *Daphnia* and *Sida* are abundant in the Bhimtal lake but are absent from Nainital lake. On the other hand *Keretalla* is found in the latter but is absent from the Bhimtal lake. This has been attributed to the difference in the levels of pollution in these two lakes.

Recession: A photograph of the Nainital lakes taken in 1921 shows the flat ground occurring on the northern shores of the lake (Mallital) to be a part of the lake. Recession of the lake waters exposed this area and what was once a part of the lake bottom is today a popular playground.

Inflow of a vast quantity of silt into the lake waters brought about by biotic pressure on the Nainital lake and its surrounds has led to this recession. If this state of affairs continues, the lake may be reduced to by about half of its present size in the not too distant future.

The Sukhatal has been virtually choked to death by inflow of sediments.

Heavy Construction

Roads connect people and places; canals provide much needed water for irrigation, drinking, washing and cooking and buildings provide shelter to mankind. In the past 30 years or so, thousands of kilometres of roads and canals and hundreds of large buildings have been constructed in Kumaun. Heavy construction leads to a number of environmental problems. These have been outlined below:

1. Removal of the vegetative cover causes problems like accelerated erosion.

2. The debris released during the excavation of roads and canals and levelling for giant buildings is thrown down the mountain slopes. It covers fields, forests pastures and may even block small streams and water channels.

3. Blasting weakens the country rocks and the whole tract is rendered prone to landslides.

4. The quantity of sediments carried by the surface run off is greatly increased.

Sewage Disposal

Sewage disposal has become a major environmental problem in all urban and semi-urban centres of Kumaun:

1. The sewage from Ramnagar is dumped into the Kosi river.

2. Nainital town dumps its sewage into the lake.

3. Haldwani dumps its sewage into a seasonal stream that flows in the eastern part of the town.

4. The sewage of Ranikhet is dumped into nearby streams.

5. Small towns like Dharchula and Pithoragarh dump their sewage into the Kali river.

Thus, most water bodies of the Kumaun hills are polluted near human settlements.

Wildlife

Due to a number of causative factors there was a steady decrease in the population of wild animals in Kumaun, particularly from 1935 to 1975. The number of tigers in the lower hills dwindled to an all time low. The same was the case with many other mammals, birds and reptiles. Fortunately for mankind, this trend has now been reversed due to intensive management and conservation of wildlife.

The causative factors responsible for damage to wildlife in Kumaun are:

1. Degradation of the forest cover which is the habitat of wild

animals. They had to face a scarcity of food and shelter.

2. Predation and competition, particularly from domestic cattle of the local people and graziers.

3. Poaching and illegal hunting in the past.

4. Diseases from domestic cattle.

5. Distrubance by tourists, particularly in the Corbett National Park area.

Natural Processes

Besides the above causes, a number of natural processes too are responsible for the degradation of the environment in Kumaun. These have been discussed in the following text:

Earthquakes: Like other parts of the Himalaya, Kumaun too lies in a seismically active belt. There have been hundreds of shocks in this region of which a few recorded ones have been destructive. Earthquakes cause the following damage:

1. Human beings and animals either perish or are injured.

2. Houses and other buildings are destroyed or seriously damaged by earthquakes.

3. Forest, pasture or agricultural lands are damaged.

4. Roads, canals and bridges are severely affected.

5. Telegraph, telephone and electric transmission lines are destroyed or damaged.

6. Earthquakes trigger off landslides, landslips, rock slides and soil slumps.

7. The course of rivers and streams may be altered or blocked by earthquakes.

A devastating earthquake occurred in the Dharchula area in December 1980 due to displacement along the main central thrust and in its offshoots. A large area of Pithoragarh district was devastated and many human beings lost their lives.

Glaciers

A glacier is a naturally moving body of large dimensions made up of crystalline ice formed on the earth's surface as a result of accumulation of snow. Many glaciers of varying sizes are found on the slopes of the main Himalayan range in Kumaun. The major glaciers of this region have already been described in a separate chapter.

Environmental problems connected with glaciers in Kumaun are outlined below:

1. Most glaciers have a negative ice balance, i.e. more ice is melting than the snow that is accumulating. This means that glaciers have been receding. The Pindari glacier has receded by about 3 kms in the past 150 years.

2. Receding glaciers leave behind unconsolidated debris which may become the source material for landslides and landslips in the near future.

3. The morainic debris is very easily erodible as it is more or less without a vegetative cover.

4. The material loosened by the moving river of ice keeps falling on the valley floor, thus offering fresh surfaces for attack and facilitating avalanching.

5. The huge areas vacated by glaciers lies barren for a considerable period of time. There is excessive erosion of morainic material during this period as the erosive capacity of the melt water is more than that of the glacier itself.

6. Moraines block the passage of the melt water and form temporary lakes. After some time this debris dam may give way, causing the accumulated water to drain away with considerable force thus leading to flash floods.

Snow Avalanches

Snow avalanches involve the quick movement or sliding down of a huge mass of snow. Avalanches normally occur in the higher Himalaya of Kumaun. They cause the following damage:

1. Human and animal lives are lost.

2. They destroy forests, pastures and even arable land.

3. Buildings, including homes are damaged or destroyed by avalanches.

4. Avalanches, particularly if they are accompanied by rock fragments scour the hill slopes. They lead to the formation of gullies particularly along weak zones. After the snow melts, these gullies are an open invitation to accelerated erosion.

River and Stream Erosion

The geological work done by rivers depends upon its kinetic energy, which in turn is proportionate to the mass of the water and the square of the stream velocity (rate of flow) or, in other words, the more the water a river carries the more is its erosive capacity, The work done by rivers and streams in the Himalaya falls into three broad categories.

Erosion: This is the most prominent feature of work of rivers in the Himalaya. Erosion may lead to the deepening of the river valley (by erosion of the river bed); broadening of the river valley (by erosion along the valley sides) or lenthening of the river valley (by headward erosion).

Transportation: The transportation of sediments may be carried out in the form of solution, suspension or siltation.

Deposition: This is brought about under the following conditions:

—when there is drop in the speed of the water

—changes in temperature and pressure conditions, and

—obstruction in the flow of the river.

River and stream erosion too is responsible for degradation of the environment:

1. The fast flowing water cuts a steep gorge. In those portions where the rocks are weak, thinly bedded or tectonically disturbed, undercutting of the valley side results. This leads to the failure of valley slopes thereby causing landslides.

2. The load of most rivers and streams consists of huge boulders and rock fragments. This debris may be deposited over fields, forests or pastures during flash floods in the monsoon season.

3. In the terai and bhabar tract, rivers and streams may change their course, thus laying waste vast tracts of good land.

4. Sots or 'raus' are seasonal streams that originate in the Siwalik hills and debouch either in the dun valleys or in the terai and bhabar tract. Their catchment consists of loose unconsolidated sediments, gullies and hogback ridges. In the rainy season a vast quantity of water pours down from the skies. The slopes of the Siwalik hills are unable to absorb it and the water flows on the surface where it reaches the sots or raus. Thus, these streams are in spate.

A large tract of land on either sides of the sots or raus is laid waste each year. It is covered by boulders of varying sizes brought down from the Siwaliks. After the monsoon season, the beds of sots or raus become dry once again and remain in this condition till the next rainy season.

12
Places of Interest

Tourism in Kumaun is today a fast expanding component of the economy. In the past only places like Nainital, Ranikhet and Baijnath were of interest to the tourists and pilgrims. Today, there are over a hundred places that are worth visiting. These places are frequented by tourists, pilgrims, sages and adventurers.

During the last two decades there has been a tremendous increase in the number of tourists arrivals in various parts of Kumaun. Tourism in Kumaun has received a major boost due to the following causes:

1. It is a quicker means of development as compared to agriculture and industry.

2. There are many places of scenic beauty and pleasant climate in these hills.

3. Religious and historical sites are common in Kumaun. They are of interest to pilgrims.

4. Wildlife areas are of interest to nature lovers.

5. Trekkers and mountaineers too come to Kumaun to climb mountains, reach glaciers and cross passes.

6. The ancient pilgrim route to Mansarovar lake in Tibet passes through Kumaun.

7. There are ample opportunities for water sports like sailing and boating in the lakes of Kumaun.

8. The Kumaun hills are well connected by air, rail and road to the major population centres of north and north west India.

9. There are ample facilities for boarding, lodging and health care in most parts of Kumaun.

Places of interest to the tourists, pilgrims, wildlife enthusiasts, adventurers, scientists, naturalists and the laymen have been discussed in brief in the following text. These include: (a) places of religious significance, (b) places of tourist interest, (c) adventurous places, and (d) places of nature and wildlife interest.

Nainital

Nainital, the headquarters of Kumaun division is the most popular place of tourist and religious significance. It is situated in the lower Himalaya not very far from the plains. This hill resort is well connected with most places in northern and north-western India. The main attractions of Nainital are:

a) It has a pleasant climate from April to October and offers relief from the scorching heat of the plains in this season.

b) The Nainital lake is one of the most beautiful lakes in western and central Himalaya (after the Dal and Nagin lakes of Kashmir). This lake provides immense opportunities for boating and sailing.

c) Nainital is also a place of religious interest. The lake basin of the Nainital is believed to be one of the places where the left eye of Sati fell when Lord Shiva was wandering in a daze carrying her corpse. Sati took rebirth as Parvati, the daughter of the Himalaya and once again became the consort of Lord Shiva.

The temple of Naina devi situated on the banks of the lake is also revered by thousands of devotees.

d) In winter there is snowfall in Nainital. This is an added attraction for tourists.

e) There are many places for trekking in and around Nainital, e.g. Tiffin top and Naini peak.

There are many hotels and motels to suit every pocket in Nainital.

Bhimtal

Bhimtal is a small town around the Bhimtal lake near Nainital. Thousands of tourists visit Bhimtal each year. The lake offers immense opportunities for boating. The hills around this town are the nature lovers' delight. Bhimtal is also a summer resort and refuge from the scorching heat of the plains. In winter its climate is mild. There are a number of good hotels in Bhimtal.

3. Bhowali

This is a mountain hamlet near Nainital. It is known for the surrounding pine clad slopes; orchards and gardens. There are small hotels in Bhowali. It is a nature lovers' paradise. The hills around Bhowali abound with birds and animals.

5. Kainchi

Kainchi is situated to the north of Nainital on the road to Almora. A famous temple stands along the banks of a small stream. It was established by Neem Karauli Baba. The temple is revered by thousands of devotees. There is an atmosphere of religiosity and spiritualism in the temple. The pilgrims flock to Kainchi from far and wide.

6. Hanuman Garhi

This is a religious complex near Nainital. The temple has a 4 mt high image of Lord Hanuman. This complex too was established by the famous saint Neem Karauli Baba. Thousands of devotees visit Hanuman Garhi each year.

7. Gorkhakhal

Gorkhakhal is a small town situated to the north-east of Nainital. It is known for its natural beauty and proximity to the plains. There is also a temple of Bhairav where many people come to worship. Gorkhakhal is fast coming up as a place for a quiet mountain holiday.

8. Nanakmatta

Nanakmatta is a holy place of Sikh pilgrimage. Guru Nanak, the founder of Sikhism visited this place. It is situated in the terai to the south-east of Nainital. Many pilgrims visit Nanakmatta every year.

Another religious place near Nanakmatta is Mitha Ritha on the banks of the Ladhiya river. Guru Nanak plucked fruits from a Ritha tree which is standing even today.

There are gurudwaras at both these places for spending the night.

9. Corbett National Park

The famous Corbett National Park lies partly in Nainital and partly in Pauri district. The administrative headquarters of this park are located at Ramnagar in the foothills of Nainital district. This park is visited by thousands of wildlife enthusiasts each year. The main attractions of the Corbett National Park are:

a) It is the home of hundreds of species of animals. The prominent ones are the tiger, leopard, elephant, spotted deer, hog deer, sambhar, barking deer, porcupine, wild boar, red jungle fowl, pheasants, quail, woodpeckers, tits, wrabblers, gharial, muggar, python, king cobra, cobra, krait and fan throated lizard. Most of these animals can be viewed in park area.

b) The reservoir formed by the Kalagarh dam is visited by many migratory birds in winter. Most of these are beautiful and attractive.

c) The waters of the Ramganga river present an attractive sight as the river flows through the Patli dun valley of the park.

d) Another attractive sight is the densely forested hills of this national park.

e) The famous temples of Garjia devi is situated on the outskirts of the park.

f) There is also a small museum at the Dhangarhi gate of the park.

Facilities for spending the night are available at Ramnagar and Dhikala.

10. Almora

Almora, the ancient capital of Kumaun is a fairly large town situated in the middle Himalaya to the north-north east of Nainital. It stands on a 1800 mt high ridge. This town is well connected with other parts of Kumaun by road. It also has easy access from the plains of northern India.

The main attractions of Almora are:

a) Its pleasant climate offers a welcome relief from the scorching heat of the plains in summer.

b) The forested hills around Almora present a scenic and enchanting view.

c) The famous Nanda Devi festival is held at Almora every year in early autumn. Thousands of pilgrims and devotees converge at Almora in this season. The image of Nanda Devi is worshipped and taken in a procession for immersion.

d) There are many temples in Almora which include the Shiva temple now known as the Nanda Devi temple; the Raghunath, Astha Bhairav, Lakshmeshwar and Badarinath temples.

There are small hotels for tourists at Almora.

11. Ranikhet

This is one of the most beautiful mountain resorts in the U.P. hills. It is situated atop a pine clad ridge to the north of Nainital. Ranikhet is well connected by road with other parts of Kumaun and the plains of north India by road.

The main attractions of Ranikhet are:

a) It has a pleasant climate.

b) There is a beautiful, breathtaking view of the Nanda Devi group of peaks from Ranikhet.

c) There are numerous places worth visiting in and around this mountain resort. This include the Chaubatia garden, Tarikhet and the famous Kalika temple.

d) The pine and oak forests around Ranikhet are the home of many interesting animal and bird species.

There are many hotels in Ranikhet where visitors can stay.

12. Kausani

Kausani is an enchanting and idyllic mountain retreat in the middle Himalaya near Almora. Thousands of visitors come to Kausani each year for a quiet holiday. The main attractions of Kausani are:

a) Its pleasant climate.

b) Spectacular panoramic views of the snow-clad Nanda Devi group of peaks.

c) Down below Kausani lies the famous Katyur valley whose beauty is unparalleled.

d) The hills around this mountain resort are covered with orchards and forests.

Small hotels and inns are available at Kausani. It can also be visited from Almora and Ranikhet on a day's visit.

13. Binsar

Binsar is an idyllic mountain paradise located near Almora. The hills around Binsar are covered with dense forests that present a beautiful sight. This area is also a wildlife sanctuary. The Binsar sanctuary abounds with hundreds of species of flora and fauna. There are also a number of places for trekking in this area.

Hundreds of tourists and wildlife enthusiasts visit Binsar every year.

14. Katarmal

A famous 900-year old Sun temple is situated at Katarmal, north west of Almora. It was constructed by the Katyuri kings. This temple has beautifully carved doors and temples some of which have been removed and taken for safekeeping to the National museum, Delhi after the presiding image of the deity was stolen from Katarmal.

14. Jageshwar

Jageshwar is situated in an enchanting valley covered with coniferous forest; to the east of Almora. A cluster of 164 temples built by the Chand rulers are located in and around Jageshwar. These temples were built in phases over two centuries and are dedicated to Lord Shiva. They are beautifully carved and their setting amdist tall deodar trees with a river flowing in the valley presents an attractive sight. The holy pool of Brahma kund for a ritual bath is nearby.

Other famous temples near Jageshwar are the Vridh Jageshwar, Uttar Vrindavan and Mirtola Ashram.

There are small hotels and dharamshalas at Jageshwar for spending the night.

16. Someshwar

Someshwar is a mountain hamlet north of Almora. It has an ancient temple of Lord Shiva that was constructed by Raja Som Chand, the founder of the Chand dynasty. Devotees come to worship at Someshwar in large numbers.

17. Baijnath

This is a small town situated on the banks of the Gomati river to the north of Almora. It was once the capital of the Katyuri kings. The main attractions of Baijnath are:

a) Old houses and temples in Baijnath have exquisitely carved doorways and windows. These carvings and sculptures are amongst the best in the western Himalaya.

b) There is an ancient temple at Baijnath which dates back to the era of the Pandavas.

c) The hills around Baijnath are covered with forests. There are beautiful terraced fields in the valley and lower slopes.

There are small hotels and dharmshalas at Baijnath.

18. Dwarahat

The temple town of Dwarahat is located to the north-west of Ranikhet. It is famous for a cluster of 55 temples of the Katyuri period. The architecture of these temples resembles that of Gujarat. Tourists can go to Dwarahat on a days' visit.

19. Dunagiri

This is a majestic mountain top in the middle Himalaya to the north-west of Almora. A famous temple of the goddess Durga stands atop this peak. Thousands of pilgrims visit this temple each year. There is a magnificent view of the snowclad main Himalayan peaks from Dunagiri. It can be approached from Ranikhet and Almora.

20. Bageshwar

Bageshwar is an ancient temple town situated to the east of Baijnath, at the confluence of the rivers Sarya and Gomati. The main attractions of Bageshwar are:

a) There is a famous ancient temple at Bageshwar dedicated to Lord Shiva.

b) Bageshwar was considered to be one of the stages on the old pilgrim trail to Mansarovar.

c) The popular Uttarayani fair is held at Bageshwar every year in January.

There are small hotels and dharmshalas at Bageshwar.

21. Chitai

The mountain hamlet of Chitai is located near Almora. There is a well known temple of the deity Goll or Golu at Chitai. He is believed to have been one of the generals of the Chand kings. Golu was accorded the status of a deity because of his valour and great deeds. People visit the temple and wish for many things.

22. Gananath

Gananath is a mountain hamlet north of Almora. The attractions in Gananath are: beautitul landscape of the surrounding hills, ancient temple of Lord Shiva and numerous interesting natural caves.

23. Champawat

Located in the centre of a broad valley in the middle Himalaya, Champawat was the capital of the rulers of Kumaun for a long time. The main attactions of Champawat are:

a) It has a pleasant climate.

b) The landscape around Champawat is very beautiful. Terraced fields cover the valley with the river snaking its way past them.

c) There are many places of historical interest in and around Champawat. This includes the ruins of the fort of the Chand rulers.

d) The exquisitely carved Baleshwar and Rataneshwar temples are suitated in this town.

Champawat is well connected by road with other parts of Kumaun. There are hotels and dharamshalas for night stay.

24. Devidhura

This is an ancient temple town situated on a high mountain to the west of Lohaghat. The main attractions of Devidhura are:

a) It has a pleasant climate.

b) Devidhura commands a magnificent view of the higher Himalayan mountains.

c) A famous temple of the goddess Varahi is located in Devidhura. Its original temple on a rock now lies in ruins. The Chand rulers used to invoke the goddess at Devidhura before embarking on important ventures. Even today devotees pay their respects at the temple.

25. Chhota Kailash or Narayan Ashram

This is a small mountain hamlet north of Pithoragarh on the route to Kailash and Mansarovar. It is located in the high mountains and commands a beautiful view of the main Himalayan mountain wall and the gorge of the Kali river. Chhota Kailash was a halting place for pilgrims to the holy Mansarovar lake.

26. Munsiari

This is a small town nestled at the base of the snowclad Himalayan chain in Pithoragarh district. Munsiari has excellent snow views and a climate that provides a refuge from the scorching heat of the plains. The mountains around this town are ideal for trekking. They are the home of hundreds of interesting species of flora and fauna. Munsiari can be approached from Pithoragarh.

27. Pindari

This is a fairly large glacier at the base of the Nanda Devi group of peaks. It is easily accessible by a trek up the enchanting Pindar valley. Thick sub-alpine and alpine forests extend for many kilometres along this valley. The Pindari glacier is very popular with tourists as it can be readily approached.

28. Milam Glacier

The Milam glacier lies at the base of the main Himalayan mountain wall. High snow clad peaks surround this glacier. The views are spectacular and enchanting. The glacier can be reached after an arduous trek from Munsiari. Thick sub-alpine and alpine forests cover the approach to this glacier.

29. Upper Darma Valley

This is a dry, high altitude valley near the border with Tibet. It has excellent views of the snow clad mountains that surround it. The upper Darma valley is suited for trekkers, mountaineers, wildlifers and nature lovers.

Appendix I. Landuse pattern in Kumaun (1982-83)
(Figures in hectares and figures in parenthesis are percentages of the reported area)

Item	Almora	Nainital	Pithoragarh	Total, Kumaun Division
Reporting area	7,24,225	6,98,309	6,24,685	20,47,219
Forest	3,94,432 (54.5)	4,02,640 (57.7)	3,30,289 (77.7)	11,27,361 (55.1)
Barren and uncultivable land	35,193 (4.9)	6,262 (0.9)	29,397 (6.9)	70,852 (3.5)
Land put to non-agricultural use	15,960	30,248	12,074	58,282
Cultivable waste	69,170 (9.6)	30,702 (4.4)	59,516 (14.0)	1,59,388 (7.8)
Permanent pasture and grazing land	50,758 (7.0)	1,239 (0.2)	72,743 (17.1)	1,24,740 (6.1)
Land under miscellaneous crops, (trees and grasses not included)	39,603	15,430	33,698	88,731
Current fallows	1,454	4,911	2,955	9,315
Other fallows	6,030	4,023	7,326	17,379
Net area sown	1,11,625 (15.4)	2,02,851 (29.0)	76,692 (18.0)	3,91,171 (19.1)
Net area sown more than once	75,682	1,34,263	62,564	2,72,509
Total cropped area	1,87,307	3,37,117	1,39,256	6,63,680
A) Kharif	1,10,947	1,87,782	76,692	3,75,421
B) Rabi	76.360	1,44,073	62,564	2,82,997
C) Zaid	..	5,081	..	5,081
D) Area prepared for cane	..	181	..	181
Total cropped area	1,87,307	3,37,117	1,39,256	6,63,680
Net area irrigated	12,063	1,38,270	6,615	1,56,948
Area irrigated more than once	11,521	67,412	5,406	84,339
Total irrigated area	23,584	2,05,682	12,021	2,41,287
Cropping intensity	167.8	166.19	181.58	169.66
Net irrigated area as % of net area sown	10.81	68.16	8.53	40.12
Total irrigated area as % of total cropped area	12.59	61.01	8.63	36.36

Source: *Seventh Five Year Plan, Hill Region,* Uttar Pradesh, Hill Development Department, Lucknow, 67-68.

Appendix II. Area and population of rural settlement in Kumaun

Development Block	Area in 1986 (km^2)	Population in 1981	Density (p/km^2)
Kumaun Division	20,852.8	19,76,307	95
Nainital District	6,685.5	8,04,262	120
Forest Localities	3,699.0	34,024	15
Jaspur Block	226.8	56,399	243
Kashipur Block	182.2	47,944	250
Bazpur Block	281.0	59,976	207
Gadarpur Block	228.1	68,771	208
Rudrapur Block	312.0	78,031	250
Sitarganj Block	344.4	86,472	260
Khatima Block	352.0	96,708	275
Ramnagar Block	144.0	45,909	319
Kotabagh Block	133.0	29,126	219
Haldwani Block	153.0	63,004	412
Bhimtal Block	94.0	30,721	327
Dhari Block	84.0	19,455	187
Okhalkanda Block	167.0	29,804	178
Ramgarh Block	142.0	28,080	198
Betalghat Block	143.0	29,838	200
Almora District	5,347.3	7,09,777	133
Forest Localities	407.3	2,674	7
Salt Block	305.0	57,481	188
Syalde Block	244.0	48,369	198
Bhikiasain Block	167.0	40,545	243
Chaukhutia Block	181.0	47,863	264
Dwarahat Block	307.0	57,109	186
Tarikhet Block	234.0	58,746	251
Hawalbagh Block	204.0	50,721	249
Lamgarha Block	313.0	39,205	184
Dhauladevi Block	323.0	52,930	164
Bhaisiachhana Block	376.0	29,736	79
Takula Block	315.0	53,971	171
Bageshwar Block	440.0	58,550	133
Garur Block	215.0	49,585	231
Kapkot Block	1,416.0	62,301	44
Pithoragarh District	8,820.0	4,62,248	52
Forest Localities	702.0	1,402	2
Champawat Block	448.0	32,854	73
Lohaghat Block	216.0	30,646	142
Pati Block	244.0	34,065	140
Barakot Block	181.0	21,132	117
Munakot Block	198.0	41,346	209

Pithoragarh Block	169.0	43,361	257
Gangolihat Block	314.0	59,812	190
Berinag Block	200.0	42,684	213
Didihat Block	209.0	32,180	154
Kanalichhina Block	352.0	42,174	120
Dharchula Block	2,964.0	41,221	14
Munsiari Block	2,623.0	39,371	15

After Anonymous, 1986.

Appendix III. Area and population of urban settlements in Kumaun

Particulars	Area in 1986 (km^2)	Population in 1981	Density (p/km^2)
Kumaun Division	182.2	4,06,876	2,233
Nainital District	108.5	3,32,261	3,062
Jaspur	4.0	21,242	
Maduadabra	5.2	3,263	
Kashipur	5.5	51,773	
Maduakherganj	2.0	4,722	
Bazpur	2.6	11,366	
Sultanpur	2.0	4,769	
Kelakhera	5.0	3,335	
Gadarpur	3.4	6,315	
Dineshpur	5.0	3,308	
Rudrapur	12.4	34,658	
Kichha	4.0	13,606	
Sitarganj	2.0	9,697	
Shaktifarm	7.6	5,190	
Khatima	8.4	8,443	
Tanakpur	1.0	8,818	
Ramnagar	2.5	26,013	
Kaladhungi	1.2	3,112	
Haldwani-Kathgodam	10.6	77,300	
Lalkuan	4.2	3,155	
Bhimtal	3.4	2,871	
Bhawali	1.3	3,212	
Nainital	14.3	26,093	
Almora District	37.7	47,596	1,262
Dwarahat	2.9	2,333	
Ranikhet	21.8	18,190	
Almora	8.0	22,705	
Bageshwar	5.0	4,368	

Pithoragarh District	37.7	27,019	751
Champawat	4.0	1,702	
Lohaghat	8.0	2,530	
Pithoragarh	16.4	17,657	
Didihat	4.3	2,044	
Dharchula	3.3	3,086	

After Anonymous, 1986.

Appendix IV. Size of rural settlements in Kumaun

Particulars	No. of settlements (in 1986)	No. of settlement inhabited (in 1981)	Population Size in 1981 (as per cent of colomn 3)					
			below 200	200-499	500 999	1000 1999	2000-4999	5000 above
Kumaun Division	7,287	6,999	55.0	30.8	10.5	3.1	0.6	0.0
Nainital District	1,839	1,806	41.5	30.7	16.9	8.6	2.2	0.1
Forest Localities	65	63	38.1	17.5	14.3	20.6	9.5	–
Jaspur Block	99	99	28.3	31.3	21.2	16.2	3.0	–
Kashipur Block	68	68	22.0	41.2	19.1	10.3	7.4	–
Bazpur Block	122	121	30.6	33.0	24.0	10.7	1.7	–
Gadarpur Block	70	70	10.0	21.4	32.9	25.7	10.0	–
Rudrapur Block	90	89	18.0	22.5	34.8	20.2	3.4	1.1
Sitarganj Block	120	120	14.1	32.2	32.5	11.7	7.5	–
Khatima Block	124	120	15.8	21.7	33.3	27.5	1.7	–
Ramnagar Block	186	186	57.0	30.1	10.8	2.1	–	–
Kotabagh Block	115	113	62.8	25.7	8.0	3.5	–	–
Haldwani Block	252	249	58.2	34.6	4.8	1.6	0.8	–
Bhimtal Block	106	102	54.9	28.4	11.8	3.9	1.0	–
Dhari Block	52	52	32.7	38.5	26.9	1.9	–	–
Okhalkanda Block	107	106	46.2	39.6	13.2	1.0	–	–
Ramgarh Block	130	122	62.3	28.7	4.9	4.1	–	–
Betalghat	133	126	52.4	36.5	11.1	–	–	–
Almora District	3,165	3,019	56.6	33.1	9.0	1.3	–	–
Forest Localities	18	16	87.6	–	6.2	6.2	–	–
Salt Block	258	254	57.1	34.6	7.9	0.4	–	–
Syalde Block	194	188	56.9	29.8	11.7	1.6	–	–
Bhikiasain Block	202	189	55.6	38.1	6.3	–	–	–
Chaukhutia Block	171	164	46.3	36.6	15.3	1.8	–	–
Dwarahat Block	212	202	42.6	42.6	12.4	2.4	–	–
Tarikhet Block	267	261	55.2	37.2	7.2	0.4	–	–
Hawalbagh Block	234	218	54.1	35.8	9.6	0.5	–	–
Lamgarha Block	207	196	65.8	27.1	6.1	0.5	–	–
Dhauladevi Block	242	236	58.1	33.5	7.2	1.2	–	–
Bhaisiachhana Block	131	127	58.3	33.9	7.1	0.7	–	–
Takula Block	231	214	56.1	29.4	12.6	1.9	–	–

Bageshwar Block	395	367	72.2	24.0	3.5	0.3	–	–
Garur Block	189	187	49.7	37.4	10.2	2.7	–	–
Kapkot Block	214	200	47.5	33.0	15.5	4.0	–	–
Pithoragarh District	2,283	2,174	63.9	27.6	7.1	1.2	0.2	–
Forest Localities	14	14	85.7	14.3	–	–	–	–
Champawat Block	232	212	72.2	25.0	2.4	0.4	–	–
Lohaghat Block	154	144	67.4	23.6	9.0	–	–	–
Pati Block	164	155	59.4	31.0	9.0	0.6	–	–
Barakot Block	103	98	57.2	36.7	7.1	–	–	–
Munakot Block	173	165	61.2	26.7	8.5	3.0	0.6	–
Pithoragarh Block	164	159	50.3	33.3	13.8	2.5	–	–
Gangolihat Block	325	303	64.7	26.7	7.6	1.0	–	–
Berinag Block	283	268	73.5	21.6	4.9	–	–	–
Didihat Block	167	165	66.1	25.4	7.3	1.2	–	–
Kanalichhina Block	207	203	61.6	33.0	4.4	1.0	–	–
Dharchula Block	74	71	29.6	40.8	14.1	9.9	5:6	–
Munsiari Block	223	217	69.0	24.0	6.5	0.5	–	–

After Anonymous, 1986.

Bibliography

Anonymous 1986—U.P. Statistics, UP. Govt., Lucknow.

Dhar, T.N. and Sastri, C., 1988—'Industrial Development in Kumaun' in *Kumaun: Land and People,* Ed. Valdiya, Nainital.

Jalal, D.S., 1988—'Kumaun: The Geographical Perspective', *ibid.*

Khanka, S.S., 1988—'Demographic Profile of Kumaun', *Ibid.*

Mittle, A.K., 1986—*British Administration in Kumaun Himalayas*, Mittal Prakashan, Delhi.

Moddie, A.D., 1985—Eco-development Guidelines for Tourism in Hill Areas, *Chea, Bull,* Nainital.

Negi, S.S., 1982—*Environmental Problems in the Himalayas*, BSMPS, Dehradun.

Negi, S.S., 1990—*A Handbook of the Himalaya*, Indus Publishing Co., New Delhi.

Negi, S.S., 1991—*Forest Types of India, Nepal and Bhutan*, PEBA, Delhi.

Negi, S.S., 1992—*A Handbook of National Parks, Sanctuaries and Biosphere Reserves of India*, Indus Publishing Co., New Delhi.

Pande, G.C., 1988—'Agrarian Scenario of Kumaun', in *Kumaun: Land and People*, Ed. Valdiya, Nainital.

Preter, S.S., 1972—*A Book of Indian Animals*, BNHS, Bombay.

Puri, G.S., 1960—*Ecology of India*, Oxford Book Co., New Delhi.

Ray Chaudri, 1960—*Soils of India*, GOI, Delhi.

Index